Whiteness in Zimbabwe

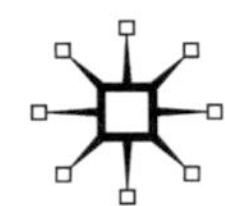

WHITENESS IN ZIMBABWE

First published in 2010 by PALGRAVE MACMILLAN® in the United States - a division of St. Martin's Press LLC, 175 Fifth Avenue, New York, NY 10010.

Where this book is distributed in the UK, Europe and the rest of the World, this is by Palgrave Macmillan, a division of Macmillan Publishers Limited, registered in England, company number 785998, of Houndmills, Basingstoke, Hampshire RG21 6XS.

Palgrave Macmillan is the global academic imprint of the above companies and has companies and representatives throughout the world.

ISBN: 978–0–230–62143–5 (paperback)
ISBN: 978–0–230–62142–8 (hardcover)

Library of Congress Cataloging-in-Publication Data

Hughes, David McDermott.
Whiteness in Zimbabwe : race, landscape, and the problem of belonging / by David McDermott Hughes.
p. cm.
Includes bibliographical references and index.
ISBN 978–0–230–62143–5 — ISBN 978–0–230–62142–8
1. Whites—Zimbabwe—History. 2. Whites—Social conditions—Zimbabwe—History. 3. Whites—Race identity—Zimbabwe—History. 4. Land settlement—Social aspects—Zimbabwe. 5. Land tenure—Social aspects—Zimbabwe. 6. Group identity—Zimbabwe—History. 7. Social isolation—Zimbabwe—History. 8. Zimbabwe—Race relations—History. I. Title.
DT2913.E87H84 2010
305.809′06891—dc22 2009028485

Design by Integra Software Services

First edition: April 2010

10 9 8 7 6 5 4 3 2 1

Printed in the United States of America.

For Melanie

Contents

List of Figures

Preface

In 2000, President Robert Mugabe and the government of Zimbabwe embarked on a project of social destruction: to extirpate the country's class of European-descended agriculturalists, known as "commercial farmers." Equal parts pogrom and land reform, the effort promised to redistribute wealth from an exclusive elite to Zimbabwe's masses. At that time, almost 4,500 white families (plus a very small number of black commercial farmers) owned 33 percent of the land area of a nation of 12 million overwhelmingly black inhabitants, and few whites showed enthusiasm for changing that state of affairs. Then, paramilitary bands occupied nearly every estate, harassed the owners, and terrorized the workers. By 2002—when the violence subsided but by no means ended—the state had removed roughly 4,000 white families from farming districts and killed ten individual whites. It had also killed large, but unverifiable, numbers of black farm workers and displaced hundreds of thousands of them. In economic terms, this disruption virtually extinguished export agriculture and jeopardized livelihoods nationwide. In psychological terms, the regime had sown insecurity in every sector of society, not least among whites, whether urban or rural. Others have written extensively about this immiseration and victimization, largely to denounce it. Mugabe's regime has defended itself with reference to the long-standing violence of racial inequality and white elitism.

This book complements those debates by focusing on the moral lives and imaginations of white Zimbabweans, both on and off commercial farms. How, before and during this period of adversity, have they understood their place in Africa, in agriculture, and under Mugabe? As conditions have turned against them, when have they adapted clear-sightedly, and when have they deluded themselves? To what extent do they subscribe to the hate and fear of racism and pigment-based prejudice? Rather little, I suggest in response to the last question. In place of such antagonism, many Zimbabweans have long practiced denial and avoidance. In their own minds, they turned away from native, African people and focused instead on African landscapes. In this nature-obsessed escape, I find commonalities with my own country, the United States, and with other European settler societies. Environmental conservation and white identity have produced and shaped each other. Beginning with Zimbabwe, this book seeks to disenthrall one structure of feeling from the other—the better to recover humanism from both.

PLAN OF THE BOOK

How have European settler societies established a sense of belonging and entitlement outside Europe? *Whiteness in Zimbabwe* seeks to answer this question in its ethnographic, comparative, and moral dimensions. European colonization, Edward Said writes, depended on "structures of feeling" wherein whites felt at home in the colonies (Said 1993:14). In the "neo-Europes" of North America and the antipodes, Anglophone whites rooted themselves in part through the genocide and expulsion of native peoples. Having attained demographic superiority, Europeans became "normal" Americans, Australians, and so on. Zimbabwe deviates from this forlorn model. After establishing the colony in the 1890s, whites never composed more than 5 percent of its population. They monopolized the land but—amid black masses—were never able to make their presence seem natural. Even as a minority, whites still aspired to belong in Africa. They could have done so by grafting themselves onto local, still vibrant, societies. To some extent, early Portuguese settlers did just that. Anglophone immigrants, by contrast, tended to adopt a strategy of escape. They avoided blacks, preferring instead to invest themselves emotionally and artistically in the environment. Of course, white farmers, industrialists, and administrators exploited blacks. But many whites chose—almost consciously—to negotiate their identity with land forms rather than social forms. To do so required extraordinary cultural work. If North Americans and Australians used violence to empty their land, Euro-Africans had to *imagine* the natives away. In what I call the "imaginative project of colonization," white writers, painters, photographers, and even farmers crafted an ideal of settler-as-nature-lover. Whiteness and conservation, in other words, coproduced each other.

The book will advance this thesis through two extended case studies: first, literature and photography representing the Zambezi River's Kariba reservoir, and second, practices of commercial agriculture east of Harare. In Chapters 2 and 3, the Kariba study delves into whites' negotiations with African landscapes. If whites preferred to compromise with the land, rather than with the people, the land still drove a hard, difficult bargain. Zimbabwe's topography differed from that of Britain in nearly every respect. Arid and landlocked, Zimbabwe contains no natural lakes and few permanent watercourses. Britain, on the other hand, has an extensive coastline and, thanks to glacial scouring, holds abundant surface water. British literature and art—epitomized by Wordsworth's description of the Lake District—values the ubiquitous, intricate boundary of land and water. As children of the glaciers and the sea, how could Anglophone settlers learn to love dry expanses of African savannah? Even the most febrile imagination could not dispel Zimbabwe's hydrological deficit altogether. If they wanted lakes, white settlers would have to engineer them, and that is precisely what the Rhodesian government did at Kariba. In the late 1950s, it dammed the Zambezi River, creating the second-largest reservoir in the world. Whites responded to the water with joy and relief, but that was not the whole of it. The new reservoir incurred high ecological costs. It flooded an enormous swath of savannah and wildlife habitat. And so, initially, white writers and photographers gave voice to a widespread sense of remorse. In the dam's second decade, however, guilt yielded gradually to Wordsworthian rapture. The beauty of the reservoir moved whites. They soon conflated beauty with nature, and, by 1977, one photographer had branded Lake Kariba a "water wilderness" for boating, fishing, and ecotourism. H_2O—once understood as the valley's misplaced molecule—eventually impressed whites as thoroughly pristine. In effect, the Kariba writers—most of whom I have interviewed—reconciled an artificial waterscape with the "myth of wild Africa." Whites Europeanized the Zambezi without losing credibility as guardians of an authentic, primeval continent. Indeed, through Kariba, Euro-Africans integrated themselves more deeply than ever into Africa's environment.

The second ethnographic case examines a similar dynamic of land, water, and belonging, but this time in the postcolonial period. Independence in 1980 and the enormity of black rule threatened to undermine whites' hard-won sense of security. By 1990, they had slipped to roughly 1 percent of national population, and the state was preparing to redistribute most of their 4,500 farms, which blanketed the fertile highveld (an Afrikaans-derived term meaning high-elevation pastures). How, under these conditions, could so few ex-Europeans feel entitled to own so much African land? To answer that question, I conducted ethnography in the Virginia area east of Harare during the 2002–03 year. In their recollections, my informants feared for their property after 1980 and, more existentially, for their place in Africa. As Chapter 4 explains, dams addressed both of these insecurities. In the 1990s, Virginia farmers impounded water for irrigation at a frenetic pace. Since government policies reimbursed farmers for improvements, expensive dams protected the farms from confiscation. As was true of Kariba, these dams also provoked a cultural response, answering whites' longing for a well-watered terrain. White farmers, in fact, referred more often to the aesthetic beauty of these dams than to their political or economic advantages. Virginia's aggregate shoreline grew by 400 percent during the 1990s. Farm owners stocked the reservoirs with bass, fished the banks, and, in some cases, used bulldozers to elongate the shoreline still further. Before politics turned against them, they had planned to stock the shorelines with wildlife and build tourism facilities. They treated the dams as nature and themselves as conservationists. Again—and without explicitly planning to do so—these farmers integrated themselves through the environment rather than through an engagement with black society. They made a hydrology of hope, blind to the gathering dangers of African politics.

Those dangers soon overwhelmed whites. In 2000, paramilitaries killed their first white farmer in Virginia, Dave Stevens. More typically, armed groups surrounded farmhouses and harassed their occupants (meanwhile assaulting and often killing black farm workers). By 2005, such tactics had removed all but 11 of the original 75 white families of Virginia. This rump, though, seems to have earned a reprieve. Chapter 5 focuses on the practical and moral dilemmas of these farmers and their less fortunate former neighbors. The farmers allowed to remain, whom I interviewed for a second time in 2005 and 2007, invoked a notion of "playing the game"—of negotiating constantly with bandits and politicians. Whites, they believed, would never hold *rights* or be considered citizens in Africa, let alone enjoy a sense of entitlement. Nonetheless—through winks, deals, and bribes—they could obtain the ephemeral *privilege* of land ownership and farming. In other words, whites in Zimbabwe have, at last, come to grips with their predicament as a minority. Virginia's remaining families focus on social, not environmental, questions. They fish less and talk with blacks more. But they do not love blacks more. In fact, the long-overdue turn to society—amid persecution—has brought fear and racial prejudice to the fore. These "post-mastery" whites grapple with an anxiety made all the more intense by its delayed onset. They have traveled from belonging to its antithesis—from the chronic hubris of settler society to the episodic terror of enclave society.

Chapter 6, the conclusion of the book, reassesses concepts of racism and conservation in light of this experience. Racism, as Albert Memmi argues, centers on "heterophobia" or fear of the other. In these terms, whites shed much of their explicit racism after Zimbabwe's independence. Expelled from politics, they concentrated on the imaginative project and on bonding themselves to African nature. Many neither feared nor loved blacks but simply tried not to think about them. They discounted the Other—a move that many blacks found more insulting than visible prejudice. The land invasions changed all that. Rural Whites either left the country or brought their imaginations to heel. They ceased to fantasize about the environment and instead engaged directly with blacks. Such encounters frustrated whites. Indeed, many describe themselves as *becoming* racist for the first time in the course of negotiating with newly militant black neighbors and politicians. Still, their actions and attitudes are bringing Zimbabwean society toward a more candid form of pluralism.

How does this sea change compare with conditions in North America, both current and desired? Here and in many of the neo-Europes, European-derived elites have made a fetish of scenery. Frontier myths situate the pilgrim or cowboy alone on empty land. Especially in the United States, an exclusionary conceit—wherein the ex-immigrant Anglo population composes the core nation—has become second nature. Majority status ensures that such entitlement will not crumble easily. Perhaps it should. Zimbabwe's predicament points to what might replace disregard for the Other: open racism of the sort one can more easily combat or, with greater luck, humanism with a cosmopolitan flair.

THE PROJECT, AS IT MELLOWED

In the 1980s and 1990s, we foreign, white academics and development workers took care to distinguish ourselves from white Zimbabweans. White Africans—regardless of their actual views—represented the old, colonial regime. Euro-Americans, by contrast, deplaned at Harare International Airport as committed nationalists and often, as in my case, as veterans of anti-apartheid student politics in the global North. As we performed this identity, we shunned our local, light-skinned counterparts. In the late 1980s, when I arrived, their company was not always pleasant in any case. A commercial farmer gave me my first lesson in biological racism, referring to Mugabe "as a fucking baboon who fell out of a tree." After that incident, I increased my distance still further. Meanwhile, I learned Shona, Zimbabwe's majority language. I conducted dissertation research in remote Ngorima Communal Land, eating *sadza* with peasants at the foot of the Chimanimani Mountains. By 1997, my informants speculated, "*Amai venyu mutema?*": Was my mother black? If, as cultural anthropologists now concur, race is an imagined, constructed category, then I stood on a threshold between two of them. A few times, I "passed" as part black.

Undertaking this project on whites, then, has cut against that grain of chosen affinities. For my dissertation and first book, I studied people whose cause I supported wholeheartedly: peasants claiming land expropriated from them. Indeed, I even helped them in small ways to regain access to such lost lands. Through "action research," I "gave back" to the community. Zimbabwe's post-2000 authoritarian turn effectively terminated this type of engaged anthropology. Paramilitary bands roamed the countryside blocking outsiders' access and harassing their interlocutors. The government suspected all foreigners—and even black Harare residents visiting communal lands—of organizing for the opposition movement. In 2002, when I made a brief visit to Ngorima, police arrested and assaulted one of my hosts. The lesson was clear: I could no longer conduct research in the black, communal lands. So, by default, I shifted to white-owned commercial farms. Of course, the state was carrying out violence there too. I could not think of approaching farm workers, let alone the paramilitary bands and politicians themselves. Yet, the commercial farmers, who were assaulted and threatened every day, would suffer no greater violence through association with me. In terms of the ethical directives of the American Anthropological

Association—to bring no harm to one's subjects—I could do the research.

Still, the question remained: how could I do the research with attention to power and inequality? The owners of large estates, attending exclusive social clubs, did not provoke the same political sympathies as did smallholders. Indeed, most previous academic work—little of which was anthropological—had denounced their economic and social position. An ethnographer, however, cannot avoid humanizing his or her subjects, some of whom, in this case, became genuine friends. More crucially, Zimbabwe's politics intervened again. At the height of my research, in 2002–03, the state was seizing much of whites' wealth. It hardly seemed necessary—or even accurate—to characterize them as latter-day landed gentry. Moreover, the confiscation of commercial farms was making peasants not richer, but distinctly poorer. Carried out with such violence and chaos, Mugabe's program of "land reform" was bankrupting every sector of the economy. Yet, even that story—of a good idea implemented badly—missed the mark. A more orderly removal of white farmers might also have worsened conditions for many rural blacks. Ironically, in exploiting black farm workers, white farm owners had also protected them. Regarding wages, for instance, commercial farmers had subtracted from the wage bill large amounts of in-kind payments, principally in the form of food. As the economy and agricultural production plummeted, those food supplies—locked in farmers' warehouses—became a political resource. Police appropriated them, even as farmers preemptively disbursed them to workers in advance of malnutrition. The dwindling population of farm workers ate better than most Zimbabweans. Perhaps, taxing wages to provide relief goods was helpful after all. Or, it was the right benefit based on the wrong reasons: farmers refused to pay full, cash wages because, as they repeatedly told me, their workers were improvident, wasteful, and would drink away the money—*as blacks would.* I found racism, sure enough—but in the service of a good cause. To put the matter bluntly, whites' prejudice-based paternalism actually sheltered blacks from a far harsher political economy. I could neither collaborate with nor oppose such a situation. Simply writing about it would be hard enough.

Therefore, I have tried to convey the moral drama of being a white Zimbabwean carrying identities and attitudes involving "race." My book follows David Roediger's *The Wages of Whiteness*—itself emulating W. E. B. DuBois—where "the tone . . . strives to be more tragic than angry" (Roediger 1991:13). For this approach, I find the notion of racism too ham-fisted: it does not capture the meanings whites associate with pigment—only some of which entail prejudice. Early in the fieldwork, I realized that many of my informants did not care very much about blacks either way. Even if they did consider their workers improvident or lazy, they did not *invest* a great deal in that judgment. They didn't care to prove it, argue it, write about it, or take photos that would demonstrate the negative assessment. They had almost, so to speak, grown weary of white supremacy. "The monologue" of insult against blacks and black leadership—which Doris Lessing detected in the 1980s—had grown quieter (Lessing 1992). In fact, many supremacists—probably including the biological racist I met in 1989—had already left Zimbabwe for South Africa, Australia, and white communities in other countries. In what, then, did the remaining whites invest their mental and emotional energy? Time and again, my interviews turned to ecology. I found men—and women to a lesser degree—obsessed with the land, soil, and water. They wrote books, shot photos and videos, and put machinery on the land in an effort to document and protect its diversity. They were conservationists first and racists second. Or, perhaps, they were not racists at all, in the strict sense of the term. Ultimately, I have striven for a new vocabulary by which to describe this intersection of white identity and ecological concern: "Other disregarding." It is an attitude of neither hating nor loving blacks, but simply not cogitating a great deal upon them. Indeed, people outside Zimbabwe—I think of the United States—sometimes turn to nature as an escape from society. In this regard, many Zimbabwean whites practiced a collective form of Thoreau-like retreat. They gained a sense of belonging, negotiated with the land and circumventing the people.

In my account, whites seek less to dominate blacks and more to find for themselves a secure place in Zimbabwe. In part, this thesis reflects the changed circumstances of a post-colony. Out of power, whites have scaled back their ambitions. Yet, belonging, I believe, always figured prominently in the white imagination. In this connection, my sympathetic approach opens up a field of comparison and contrast between white Zimbabweans and other neo-European groups. We all—I include myself as a Euro-American—struggle to make a "New World" meaningful and normal to us. We struggle to naturalize our own presence on it. In Zimbabwe, whites' obvious minority status makes the need for comfort all the more pressing. When I conducted fieldwork in Chimanimani, whites of the district consulted me on the meaning of that word. They thought it described the shape of the mountain range, but they were not sure how. They felt they needed to know, and surely an anthropologist could find the truth. In the United States—where the white majority has conferred native status upon itself—few value knowledge of things aboriginal. We forget even that such matters exist. Born in Massachusetts, I grew up only wondering why the state's name was so hard to spell. It never occurred to me that *my* language might be the foreign import. Perhaps we transatlantic Anglos could use a dose of the kind of self-doubt so palpable on the white highlands. On the one hand, I find the ambiguity and uncertainty of Euro-African culture to be salutary and worthy of emulation. On the other, the Euro-African response to that uncertainty—a turn away from natives and toward nature—is not. White Zimbabweans, I argue, misjudged their conditions. If this book is successful, it will challenge them, white Americans, and many others to think differently about conservation and nature loving—to stop ignoring social problems by romancing the land.

ACKNOWLEDGEMENTS

The people who contributed most to this work—through their hospitality, trust, and friendship—are still in Virginia, Zimbabwe. Farmers welcomed me into their homes even as they contemplated losing them forever. For their own protection, the names of these individuals must remain confidential. (I have also excluded from the book any photographs of them.) Among those informants who have left Zimbabwe, I am particularly grateful to John Gordon Davis, Steven Edwards, Denis Phocas, Steven Pratt, and Peter Robart-Morgan. At the University of Zimbabwe, the Department of Economic History hosted me during the crucial 2002–03 year. Thank you, Eira Kramer and Joseph Mtisi, for your assistance and friendship. In addition to Eira and Joe, colleagues near and far commented on all or part of the manuscript: Lincoln Addison, Jean Comaroff, Vupenyu Dzingirai, Daniel Goldstein, Amanda Hammar, Angelique Haugerud, Nancy Jacobs, Diana Jeater, Walton Johnson, Pauline Peters, Edward Ramsamy, James Scott, Julie Livingston, Melanie McDermott, Richard Schroeder, Sandra Swart, and my dedicated and discerning mother, Judith Hughes. As series editors, Martin Murray and Garth Myers provided especially helpful guidance. For technical advice regarding dams, I thank Adel

Shirmohammadi (at the University of Maryland's Department of Bioresource Engineering) and Julie van der Bliek (of the International Water Management Institute, Colombo, Sri Lanka). Leslie Wood and Lincoln Addison provided critical assistance with Chapters 4 and 5, respectively. Mike Siegel finalized the figures. Under the leadership of the late, sorely missed Glen Elder, the Northeast Workshop on Southern Africa furnished a forum and feedback for much of the manuscript. Keith Wailoo and the Center for Race and Ethnicity did me the same favor at Rutgers. At Palgrave Macmillan, I am grateful for Luba Ostashevsky's exceptionally efficient editing and for the two astute readers she recruited. On the financial side, the United States Agency for International Development supported my early research (through Michael Roth and the Land Tenure Center at the University of Wisconsin), and the Mellon Foundation made the final stages possible (through a New Directions Fellowship and supplementary funds). In 2006, an earlier version of Chapter 4 appeared in the journal *American Ethnologist* (Vol. 33, No. 2, pp. 269–287), and the *Journal of Southern African Studies* published material now found in Chapters 2 and 3 (Vol. 32, No. 4, pp. 823–838). Finally, I thank Melanie, Jesse, and Sophia for sharing in and sacrificing for this book in ways big and small.

CHAPTER 1

THE ART OF BELONGING

> Their frontier became a heaven and the continent consumed them . . . And they can never write the landscapes out of their system.
>
> —Breyten Breytenbach, *The Memory of Birds in Times of Revolution* (1996:108)

> Certain places seem to exist mainly because someone has written about them. Kilimanjaro belongs to Ernest Hemingway.
>
> —Joan Didion, *The White Album* (1979:146)

Imperial colonizers do not seize land with guns and plows alone. In order to keep it, especially after imperial dissolution, settlers must establish a credible sense of entitlement. They must propagate the conviction that they belong on the land they have just settled. At the very least—and this may be difficult enough—settlers must convince themselves of their fit with the landscape of settlement. In other words, while *excluding* natives from power, from wealth, and from territory, overseas pioneers must find a way to *include* themselves in new lands. Two factors interfere with such public and private persuasion: pre-existing peoples and the land itself. Known as natives, Indians, aboriginals, and so on, the people *settled upon* clearly hold a stronger claim to belonging. If colonization requires a contest of ancestral ties, then colonials will surely lose. The landscape itself also competes—in an oblique fashion—with settlers. As they seek to understand, name, domesticate, and farm the outback, the bush, and the desert scrub, those strange ecosystems spring traps and surprises. On riverless expanses, for example, the frontiersman finds he can neither till the soil nor mark a boundary. Amid failed crops, doubt and ambivalence overwhelm the hubris of settlement. White African writers have taken this uncertainty as their imaginative terrain. "[L]iterature," as Edward Said argues, "participat[es] in Europe's overseas expansion and . . . creates . . . 'structures of feeling' that support, elaborate, and consolidate the practice of empire."[1] Or—in the more quotidian project of settlement—writers muddle through, alternately promoting and questioning their central conceit. "All white African literature," writes Doris Lessing, "is the literature of exile, not from Europe but from Africa."[2] Taken as a whole, the settler canon has confronted that exile and persistently dampened its effects. To European and North American readers, the very existence of white writers signals the maturity and entrenchment of settler classes. To settlers themselves, fiction, memoir, and amateur history establish cultural authority—whites' capacity to understand and represent the land they inhabit. By writing and in writing, then, extra-European whites have forged senses of belonging more enduring and resilient than empire.

This imaginative project unfolded alongside and in tension with various political endeavors of the colonies and their successors. Nineteenth- and twentieth-century empires sought to exploit or uplift native peoples and often to do both at the same time. Entire professions—missionary, colonial officer, agricultural extension agent, and, arguably, anthropologist—arose in the course of this fundamentally social engagement. Although often antagonistic to one another, these specialists deformed, undermined, and reshaped African societies in fundamental ways (Comaroff 1989; Cooper and Stoler 1989; Stoler 2004). My earlier book addressed one facet of these changes: the onset of "cadastral politics" in eastern Zimbabwe, wherein whites took land from blacks and blacks made land a crux of contestation (Hughes 2006). At the risk of vastly simplifying these understandings, I refer to this entire complex of white-on-black power as an "administrative project."[3] *Whiteness in Zimbabwe* will explore what one might call a separate, parallel universe of emotion and expression: the project of belonging.[4] Far from Europe, settler colonials wanted a home. Karen Blixen, for instance, found one on the savannah. "I had a farm in Africa," she writes,

> . . . [where] the views were immensely wide. Everything that you saw made for greatness and freedom, and unequalled nobility . . . you

> woke up in the morning and thought: Here I am, where I ought to be. (Dinesen 1937:3–4)

Flying over the highlands—with her lover, Denys Finch Hatton—gave Dinesen the same feeling of attachment and completion. Settlement, then, depended not only on seizing resources from Africans but also on establishing a more personal form of ownership. Even while colonials exploited Africans, they also *dwelled* in Africa.

These two aspects of the white intrusion ran through communities and individual lives—in ways that many whites themselves sometimes recognized and criticized. After the dawn reverie, Dinesen bossed her fractious Kikuyu workforce, doing her part for administration. Rob Nixon, who grew up in South Africa's Karoo in the 1960s, describes his father as similarly split. Journalism was his job, naturalism his passion. If reportage brought him into contact with nonwhites, plants restored the distance: "[A]n amateur botanist[, . . . h]e believed you had a moral obligation to know the place you lived in, preferably in Latin" (Nixon 1999:14). In this oddly otherworldly world, Nixon's father found his true self: "That's where he composed himself: behind the camera, in front of the foliage" (ibid.:4). Under white rule, such hobbies mostly passed below the radar of criticism. Black activists and white liberals alike targeted administration rather than leisure. Nadine Gordimer was among the few white anti-apartheid writers to indict nature-loving. Her 1972 novel *The Conservationist* concerns a man "in love with his farm." The farm seems to reciprocate. "If you walk about this place on your own," the proprietor explains to his liberal activist girlfriend, "you see things you'd never see otherwise. Birds and animals—everything accepts you. But if you have [black] people tramping all over the place . . ." (Gordimer 1972:176). Notwithstanding—or perhaps because of—the police massacre at Sharpeville, the character withdraws from South African society. While striving for a more just administration, Gordimer also drew attention to the danger in white imaginings of belonging.

After and outside white rule, a range of Euro-African authors depicted these cultural dynamics with increasing clarity and bitterness. Rob Nixon exiled himself to the United States in 1980. Nearly 20 years later, he recalls in print, "after my fall into politics, the landscape around me seemed illusory . . . The Karoo became code for a long hallucination." Unlike his father, he could not reconcile wilderness and current events: "My appreciation for the bird world had long since been bankrupted by politics. Nature shrank: it seemed unnatural" (Nixon 1999:102, 104). That sense of unreality and discovered reality pervades the contemporaneous work of Italian-born Francesca Marciano. Dissecting the author's own coterie of white Kenyans and expatriates, her main protagonist rants: "For the majority of people, *whites*, I mean, the whole point of living in this country is to avoid the sight of other human beings . . . That's the whole point of going on safari, isn't it?" (Marciano 1999:194; emphasis in orginal). Some whites did not go on safari. The avant-guarde of white Nairobi, Johannesburg, and Harare socialized and even married with blacks. This interracial set contributed disproportionately to literature—but as authors, rather than as characters. Whether critical or celebratory, the canon of Euro-African writing continues to center on rural life, entangled with landforms and biota. As Breyten Breytenbach suggests in the first epigraph of this chapter, white Africans' obsession with landscape slides toward pathology. It has contributed to the delusion—shared by Joan Didion in the second epigraph—that whites should own the landscape. Sometimes indicted and sometimes abetted by the arts, the vision of white belonging in Africa shaded into claims that Africa, in fact, belonged to whites.

Nowhere was this set of fantasies and desires more deep rooted and prolonged than in Zimbabwe. Known before independence as Rhodesia, the territory suffered from a mismatch of politics and population.[5] After an uprising in 1896–97, whites pacified native polities, monopolizing politics and the economy as settlers had done in the United States and Australia. Yet they immigrated in numbers far smaller than on those frontiers. White enclaves never exceeded 5 percent of Rhodesia's population.[6] Minority status led to a curious myopia. Whereas Anglo-Americans of the eastern seaboard began to romanticize Indians in the nineteenth century—when most were safely exterminated or expelled—Anglo-Africans preferred not to dwell on the native masses surrounding them (Lepore 1998). And they chose not to dwell *with* them either. In contrast to French, Portuguese, or Dutch administrators, British colonial officers sought to prohibit rather than shape social and sexual intercourse (Rabinow 1989:294; Stoler 2004). In the early twentieth century, Rhodesians feared the "black peril" of African masculinity and miscegenation.[7] But this public hysteria subsided by the 1930s. Thereafter, informal policing kept intermarriage—and even the learning of African languages—to a minimum (Jeater 2001). Whites differed, of course, by national origin, date of arrival, and place of residence. But, through "self-serving osmosis," they identified with pioneers and farmers. An Italian, say, immigrating to suburban Rhodesia after World War II—the period of major influx—soon boasted of " 'everything we have built up here' " (Caute 1983:88). As such statements suggested, whites often overlooked the contributions of black workers. Some homeowners did establish enduring bonds with their domestics, but others treated them like the furniture (cf. Hansen 1989). "Not only had I not seen what was going on in my own home," recalls Zimbabwean Lauren St. John in a memoir of her pre-independence childhood,

> but I'd lived alongside or been in very close proximity to Africans all my life and yet we'd led completely separate lives. What did I really know of them and their struggles beyond a checklist of generalizations? (St. John 2007:203)

Many whites, in other words, swam in the ocean of social knowledge without getting wet.

In this context of self-imposed isolation, Rhodesians adapted to Africa through their imaginative project. They did so on broadly environmental terms. By fixating on the land, white writers and their readers put out of their minds the *social* exile in which they lived. No essential law dictated a zero-sum game, but myopia toward society correlated with 20/20 vision for nature. "I was seeing, really noticing, Africa in a way I'd never done before," recalls St. John upon her move to Rainbow's End farm in 1978, "the way the bush turned a deep dense black long before the sky ever did . . ." Such romance did not remain innocent. St. John's familiarity with nature and her ignorance of people converge on a property claim:

> . . . how deliciously remote Rainbow's End was. As if no one existed but us. That already it felt like home, like *our* farm, and that the future stretched like an unending road ahead of us, electric with promise (St. John 2007:114; emphasis in original)

The guerrilla war, raging on their doorstep until 1979, hardly dented the family's insouciance. Particularly among rural farmers, literature and memory enfolded whites in an echo chamber of environmental narration.

Even after Zimbabwe's independence in 1980, many whites continued to avoid a full reckoning with society and their own minority status. In the cities, their previously all-white neighborhoods integrated gradually and then more rapidly as the 1990s' policy of economic structural adjustment enriched black elites. But few whites placed themselves in contexts where blacks outnumbered them. Many liberals welcomed blacks into their restaurants and clubs. Fewer crossed the color bar themselves to attend, say, a football match in a township stadium. Fewer still learned Shona, Ndebele, or any other of the country's Bantu languages. And—where they obtained—these experiments remained surprisingly private: white writers hardly mentioned them. Indeed, white authors and, by extension, their readers continued to forge their identity outside the city and the built environment altogether. As the Harare conservationist Dick Pitman writes, "rhino fever" and similar forms of animal-love answered the needs of "what was rapidly becoming a disempowered ethnic minority looking for a role" (Pitman 2008:106). Euro-Zimbabweans still lead many local conservation NGOs, dominate the ecotourism business, and publish text and photos related to these activities (Uusihakala 1999:37). As a calling, nature conveys a moral force of greater universality than evanescent politics. The bush will outlast Robert Mugabe and land reform. Whites' symbolic kinship with plants and animals, thus, helps to naturalize lighter pigments in Africa. "Race and nature," as Moore, Pandian, and Kosek (2003:1) argue, "work as a terrain of power." Through art and expression, Zimbabwean whites have alternately maintained power and maintained themselves against power.

Escaping African People

For European overseas settlers, bilateral human-land relationships frequently emerged from more complex triangular systems. In a comparison of the United States and South Africa, for instance, George Fredrickson emphasizes a process involving three terms: *colonizers* "struggle with the *original occupants* for possession of the *land*" (Fredrickson 1981:4; emphasis added). Before genocide decimated the original occupants, Euro-Americans *did* engage with them. In the Great Lakes region, seventeenth-century settlers and Indians established a "middle ground" of shared politics, kin networks, and even religion (R. White 1991). Some white Africans did the same. The earliest—one might say, proto-imperial—settlers intermarried. The

Portuguese *prazeiros* of sixteenth-century Mozambique ascended to local chiefships, eventually losing all European ties (Isaacman and Isaacman 1975). Much later, colonial governments imposed a strict racial order, segregating blacks into rural reserves and urban townships and reducing intercultural contact to a minimum. Still, they could not segregate history and meanings. In many rural areas amid and around white settlement, memories of natives—and often natives themselves—litter the landscape. Upon entering the native reserve, the girl in one of Lessing's stories experiences "meaningless terror" and senses "a queer hostility in the landscape . . . it seemed to say to me: you walk here as a destroyer" of African society (Lessing 1951:56, 58). Africans, as the third point of the triangular relationship, would not go away. With effort, though, writers more loyal to the endeavor of settlement than Lessing could minimize and ignore them.

Mainstream Rhodesian writers crafted a property claim and self-image around the figure of an absent native unworthy of his environment.[8] The discourse of European conquest already provided rich precedents (Gordon 1989:147). In New England, for example, seventeenth-century Puritans considered the landscape to be utterly wild. Indians lived there, but they were not cultivating or "improving" fixed, fenced parcels (Cronon 1983:57). "[T]he tribes didn't actually inhabit the land. Rather they ranged over it," paraphrases Henry Reynolds, in a critique of nineteenth-century Australian rhetoric (Reynolds 1998:20). In Rhodesia, settler fiction *did* acknowledge some natives, notably the "noble savages" of the Ndebele minority. But when confronted with evidence of true civilization—such as the Great Zimbabwe ruins—writers credited Phoenicians, Arabs, and other long-departed semi-Caucasians (Fontein 2006:3–18). Shona, who constituted the bulk of Rhodesia's black population, could not possibly have built stone structures. Apparently, they debased the very terrain. "[T]his wide lovely land calls for some nobler destiny than to be the necropolis of the wretched Mashona nation," declares Cynthia Stockley in the first settler novel. "It is a white man's country."[9] This rhetoric of "Rhodesian pastoral"—as Chennells (1982) argues in his masterful history of settler fiction—cast Europeans as uniquely capable of appreciating, enhancing, and glorifying the environment. "I *love* the world—the *original* world—and I think there is only one thing worth doing for a man and that's to preserve it," writes Michael Fisher in *The Dam* (1973:42; emphasis in original). For reasons that will be explored in Chapter 2, he pursues this goal by impounding rivers. The land repays the favor: "No longer was he a small freckled man, he was huge, a giant, there was no limit to his vigour" (Ibid.:226). And the mutual seduction continued. The land is "lying up there like a woman . . . ," wrote W. A. Ballanger of the pioneer moment, "waiting for the first man to come along and take her" (Ballinger 1966:16). In this highly gendered structure of feeling, whites consummated their bond with "virgin land."[10] Blacks—if they entered the text at all—shifted around perversely and impotently. Literature thus simplified Africa, sealing its people into two air-tight groups and then largely forgetting one of them.

This process of self-blinding progresses, step-by-step, in Dick Pitman's personal story of immigration, *You Must Be New around Here*. Pitman, who later became a leading conservationist, left England in 1977, seeking to escape the everyday. So primed, he immediately noticed social difference in Salisbury: "there were, I thought, rather obviously, a lot of black people around . . . I treated every African like a kind of bomb; gingerly, in case he exploded at me in a frenzy of racial fury" (Pitman 1979:5). Such hazards would have sent many a more faint-hearted newcomer into full retreat. Indeed, as Pitman well understood, white domination had driven some blacks not only into a rage but into a guerrilla struggle, at that point four years old. Still, Pitman continues his cross-racial exploration. "I decided I would learn Shona," he recalls. "I was getting frustrated at not being able to communicate with nine tenths of the population" (Ibid.:10). Abruptly, however, the landscape intervenes and derails Pitman's cultural curiosity for good. Driving out of the capital—as the vista widens—he realizes how England had "squeezed [him] sideways into the tiny interstices between towns and villages . . . [I] had—in Africa—suddenly ballooned up to full life-size" (Ibid.:28). Pitman not only discovers himself but discovers—or invents—a new Rhodesia. After three days in the mountains of Nyanga, he describes "a feeling that I have often since experienced in the wild places: this cannot be the same country that is being slowly bled to death by a terrorist war" (Ibid.:31). Such denial of reality appears periodically in the rest of the book. At Lake Kariba, "the war has never pervaded or altered the beauty of these wild, deserted, and timeless areas. Maybe Africa itself, the real Africa, is too eternal and timeless to care" (Ibid.:49). On a different visit to Lake Kariba, "I rediscovered Africa in all the intense strangeness and fascination that had been swamped by reaction to people and events" (Ibid.:130). Finally, Pitman opts to settle in Rhodesia, or as he puts it, "the magic of Africa, of the vast spaces; the silences; the impassive, eternal panoramas of bush, mountain, rock and river was taking hold" (Ibid.:173). What is astonishing for so momentous a decision is how little it relied upon observation. While covering the war as a journalist—and actually serving in the army—Pitman consciously constructed an alternate universe, the history-less continent unaffected by human sorrows and triumphs. Ultimately, Pitman immigrated twice: first, to a Rhodesia in black-white, and second to an "Africa" of the mind.

For most whites, however, war redirected consciousness toward the social world. Increasingly evident from 1973 onward, black men wielding AK-47s inspired hate and fear in and out of print (Frederikse 1982:167ff). They targeted white farmers and soldiers, whose draftee ranks swelled to include all men under 50. Novelists depicted actual, fictional, and ambiguous instances of black-on-white mayhem. Peter Armstrong's macho *Operation Zambezi* (1979), for instance, begins with the infamous Viscount attack, in which guerillas allegedly shot down a Kariba-bound passenger plane and executed survivors on the ground. Worse still, as Armstrong insinuates, "There was evidence to suggest that female survivors had been raped" (Armstrong 1979:5). Such incidents—whether real or daydreamed—cut through escapism, fixing white imaginations on blacks and even on the black peril. Not all these emotions were negative. Written and oral accounts attest to the camaraderie and the deepest of loyalties between black and white soldiers serving the state's army in remote, rural locations. Indeed, in an effort to blame abuses on guerrillas, white soldiers of the Selous Scout regiment disguised themselves as blacks and prided themselves on the most convincing mimicry (L. White 2004). In this single-sex, explicitly nonsocial space, white men could and did establish intimate bonds across the racial bar.[11] Still, the empty landscape persisted in near hallucinations. Journalist David Caute captures the mood when, visiting a besieged farm in 1976, he converses with the heavily inebriated woman of the house: "She rambles on, nostalgic, about the old days when you could picnic anywhere, or ride on horseback cross-country, camping and fishing up at Inyanga, boating on Lake Kariba . . ." (Caute 1983:34). In *vino*, even the most embattled Rhodesians still expressed their own eco-*veritas*.

When whites lost the war, almost Orwellian mechanics for forgetting swung into full operation. Nearly three-quarters of the white population emigrated between 1979 and 1990. The plurality went to South Africa, where many whites had responded differently to their own race-based horrors. Pretoria's massacre of unarmed demonstrators in Sharpeville in 1960 compelled writers' attention. "It began to be apparent," recalls J. M. Coetzee, "that the ultimate fate of whites was going to depend a great deal more urgently on an accommodation with black South Africans than on an accommodation with the South African landscape" (Coetzee 1988:8; Foster 2008:2). North of the Limpopo, however, postwar whites more frequently sidestepped such a reckoning. An ex-soldier relates to Vincent Crapanzano, "After a while you begin to feel the call of the bush again . . . You forget the horror of it. You remember the beauty . . . the sunrises, the stars, the smell" (Crapanzano 1994:882). Authors appear to have followed the soldier's lead. As the editor of a 1982 poetry collection noted with approval: "[D]espite the years of war and upheaval, and the participation or involvement of the writers in it, their preoccupation is very much with the mundane, and with Nature and the seasons of the land they loved" (Bolze 1982:x). There is no metaphor or subtext here. African nature, in this literature, does not symbolize African people, as it does, for instance, in Gordimer's *The Conservationist*, where drought and flood destroy the white-owned farm (cf. Coetzee 1988:8; Nuttall 1996). Nature stands for itself, as one of the contributors to the 1982 collection makes plain. In "I and the Black Poet," John Eppel contrasts himself—savoring "a memory of crocus bulbs"—and his counterpart: "He focuses on Sharpeville and Soweto" (Eppel 2005:40–42). That poem first appeared in the 1970s. By 2007, Eppel had written a handful of novels centering on political and economic corruption in Bulawayo (e.g., 2002 and 2006), but his poetry still fetishized crocuses, the Matopos hills, and so on. The choice of subject was deliberate. Eppel was, as he explained to a literary magazine, trying "to find a voice which merges British form (prosody) with African content (mostly nature)" (*Leicester Review of Books* 2007). Blacks—evidently capable of killing whites and being killed by them—still did not seem to rank as publishable "content."

Or, when they appear in print, blacks seem to speak for the environment, rather than for themselves. Alexandra Fuller's widely acclaimed memoir, *Don't Let's Go to the Dogs Tonight* (2001), recounts a confused childhood during and shortly after the war. At a mixed-race school, blacks laugh at her sunburn. To Fuller, this humiliation stems from climatological, as much as social, difference. "My God,

I am the *wrong* color," she recalls. More specifically—as if marooned in the torrid zone *sans* pith helmet:

> The way I am burned by the sun, scorched by flinging sand, prickled by heat. The way my skin erupts in miniature volcanoes of protest in the presence of tsetse flies, mosquitoes, ticks. The way I stand out against the khaki bush like a large marshmallow to a gook with a gun. White. African. White-African. (Fuller 2001:10, emphasis in original)

Her alcoholic mother, who "has lived in Africa all but three years of her life," muddles the matter further: " 'But my heart'—Mum attempts to thump her chest—'is Scottish' " (Ibid.:11). Fuller's own heart attaches itself to Devuli Ranch, a dust-dry section of the lowveld marked on maps as "Not Fit for White Man's Habitation" (Ibid.:161). The family herds cattle over "this landscape which is turn-around-the-same no matter which way you face," and—stranded in a remote section of bush—Fuller nearly dies of dysentery (Ibid.:164). Recovery seems to forge her identity, or at least she "make[s] a vow never to leave Africa" (Ibid.:179). To reside on the continent, then, was to struggle with the land and survive (Harris 2005:114–117). Fuller also encounters black squatters and, especially when the family moved to Malawi, an abundance of black neighbors. Yet, these interruptions by black society only delay the family's ongoing project of environmental home-finding. *Don't Let's Go to the Dogs Tonight* ends with Fuller's parents settled on the north bank of the Zambezi. "Chirundu," she opines, "is one of the least healthy, most malarial, hot, disagreeable places in Zambia. But it is, as Dad says, 'far from the maddening crowd' " (Ibid.:299). Among the available forms of madness, the Fullers chose the constant peril of drought and disease.

Did such almost willful social avoidance and indifference rise to the level of racism? Yes, but only if the definition of racism expands voluminously. Linguistic anthropologist Jane Hill, for example, considers English speakers' use of "mock Spanish" be a "covert racist discourse" (Hill 1999). "I will do it *mañana*" indexes a lazy Mexican native and elevates North American whiteness. Hill's analysis is persuasive, but it gains little from the added charge of racism. Like Albert Memmi (2000), I would rather reserve the category of racism for a narrower set of sentiments, centering on an explicit fear of and hatred toward the Other. Or perhaps centering on less emotional sentiments: George Fredrickson, comparing the United States and South Africa, associates racism with the claim that differences of culture, status, and power "are due mainly to immutable genetic forces" (Fredrickson 1981:xii). More recent scholarship acknowledges a shift from biological to cultural prejudice, wherein the racist attributes the same determinative function to history and heritage (Holt 2000:13–14). Whether agitated or contemplative, then, this archetypal racist takes an interest in the qualities and doings of those who are different. The Other is present in racist thought. In this sense, colonial whites' administrative project certainly advanced with racist intent and outcomes. Rhodesian segregation and labor management, like apartheid, arose from "an essential racism in which people of color are considered to be *quintessentially* different from whites . . ." (Crapanzano 1986:39, emphasis in original). Face-to-face, Rhodesian farmer Basil Rowlands could kick his laborer to death in the late 1970s and leave court with a fine equivalent to US$600 (Caute 1983:132; Moore-King 1988:122). The incident conjoined Rowlands's personal animus with the judiciary's structural prejudice. A narrow definition of racism still describes much of Rhodesian history and Zimbabwean life.

But not all of it: before work and on weekends, even Rowlands might shift into a different frame of mind—into the imaginative project. Fishing by the dam, for instance, entailed neither a posture of hating nor an openness to loving blacks. For the moment, one might simply not care much about them at all. This consciousness disregards the Other. "Did we talk about the Africans?—the blacks—the 'munts'—the 'kaffirs'?" asks Lessing in her autobiography, dabbling herself in the racist vernacular. "Not much," she answers (Lessing 1994:113; cf. Chennells 2005:142). Blacks bulked small. In comparison, the land, plants, and animals bulked large.[12] "They come [to the Matopos] to picnic, fish, catch butterflies, and photograph the game," wrote the travel writer Evelyn Waugh (1960:141). "Most Rhodesians seem to me morbidly incurious about native customs and belief." Often, they did not *see* blacks, even if the latter outnumbered most game species. This subtle form of exclusion aroused little attention—especially in comparison with the violent exclusion practiced all around it. Even among scholars, outrage at an atrocity obscures the analysis of secondary problems. Perhaps for this reason whites' safari mentality has persisted almost uncriticized into the era of black enfranchisement and continent-wide prominence.

In Zimbabwe, whites may no longer kill a black with impunity, but they can still lose themselves in the bush. Until very recently, this self-segregation allowed those with light skins to escape both the progress of racial integration and the dead-end of racial animosity. The evasion—even if it did not constitute racism—still hurt both black and white Africans.

CHILDREN OF THE GLACIERS

From the pioneer days of the 1890s, Rhodesia, like Kenya, presented unprecedented environmental challenges. Theories of "non-cosmopolitanism" warned bluntly that Northern people would die in the tropics, known as the "torrid zone" (Redfield 2000:192–196). "[T]the white man must be content to settle there temporarily [and] to teach the natives the dignity of labour," admonished E. J. Ravenstein, a leading proponent of non-cosmopolitanism. Yet, in 1890, Ravenstein admitted to "an exception to the rule": tropical uplands (Ravenstein 1891:31; cf. Bell 1993:331). With cooperation from the climate, altitude could mitigate the effect of latitude. That very year, the British South Africa Company sponsored a pioneer column to advance north from South Africa into what is now Zimbabwe's highveld. In the same decade, British settlers established Kenya's white highlands. Above the 1,000-meter contour in Zimbabwe and above 1,500 meters in Kenya, whites could and did put down roots. Still, fear persisted. The sun—among many threats from nature—seemed to beat down on settlers' heads. Pioneers feared skin damage and worse: solar-induced sterility (Kennedy 1987:115). Those settlers who could not avoid exposure to the sky wore pith helmets and clothed themselves from head to toe. "No medieval knight could have been more closely armoured . . . against the rays of the sun," writes Elspeth Huxley (1959:7), in a memoir of her childhood on Kenyan farms in the 1910s and 1920s.[13] By Huxley's adulthood, high survival rates had dispelled these anxieties. Still, the sense of quarantines imposed and lifted fixed whites' attention on medical geography. Huxley, who became the literary voice of white Kenya, later reflected, "I do not think it occurred to anyone that politics, not health, would decide the issue" of whether it was "a white man's country" (Huxley 1985:54; cf. Huxley 1967).

At a conceptual level, this very association between land and people shaped whites' African experience. European migrants, particularly those from the British Isles, packed their sea trunks with an ambiguous environmental heritage. Since at least the Enlightenment, they had treated their surrounds as purely material. Modern rationality disenchanted forest and mountain alike, reducing land to the status of a useful object (Glacken 1967:462–463). That utilitarian stance equipped Europeans quite well for travel: long-distance movement only implied a change of practical context—substituting one agrarian system for another—not a reorganization of self or values. Even the value-laden dictates of Christianity facilitated relocation. Portable, Latin blessings could render any water—not merely that of known sacred springs—suitable for the sacraments. In Word and worldview, then, Europeans built themselves for mobility. Their conscience, as Pratt (1992:15ff) demonstrates, was "planetary." At the same time, learned ideas could not altogether overrule the accumulated weight of lived experience. Residence in Britain had imprinted Britons with an affinity for British landscapes. And, fortuitously, those landscapes differed markedly from much of the rest of the world: glaciers had scoured and molded them, a past that northern Eurasia shared only with a roughly parallel band of the Americas.[14] Such topography did not determine white attitudes or actions. An intertwinement with this environmental history did, however, equip whites rather better for staying at home than for traveling, especially to the tropics. In moving south, they would inevitably cross a profound divide. English settlement in Africa has unfolded as a contest between these two forms of heritage: the engine of whites' capacity to adapt to new geographies against the brakes of whites' geographical custom.

In large part, that custom or sensibility depended upon water and other forms of moisture. Above all other factors, ice and temperate rainfall produced long and intricate boundaries between land and water.[15] Glaciers, melting for the last time 12,000 years ago, sent water in channels to the coast, indenting it at frequent intervals. "The kingdom by the sea"—in Paul Theroux's (1983) phrase for Britain—weaves in and out of the Atlantic from peninsula to inlet to spit. This baroque curvature continues inland as well. Glacial meltwater unable to reach the sea formed an inverse archipelago of interior lakes, and year-round rainfall has kept them full. Also full of meaning: lakes feature in a body of European literature far too vast to summarize here. In English, perhaps no author has described a lacustrine landscape with greater directness, precision, and confidence than William Wordsworth.[16] "It is much more desirable, for the purposes of pleasure," Wordsworth opines in an 1822 guide to the Lake District, "that lakes should be numerous and small or middle-sized than large, not only for communication by walks and rides, but for variety." The optimal shoreline, he continues, "is also for the most part gracefully or boldly indented" (Wordsworth 1991[1822]:22–23). Such an intricate boundary of land and water added variation and interest to a landscape. In the previous century, Edmund Burke had already established a notion of *beauty* in landscape painting based on moderation—as opposed to the *sublime* or terrifying (Nicolson 1959:313). Cresting in the nineteenth century, the Romantic Movement delved deeper still into these connections between physical geography and personal emotion, between gardens and passions (Thomas 1983:262). In this context, then, Wordsworth offered an exact formula: the alternation between wet and dry, cove and peninsula, blue and green set the eye and the spirit at rest. Beauty found its geometry.

The tropics violated this standard in every way. After a period of Iberian conquest and monopoly, Northern European explorers began to visit and study Central and South America in the eighteenth century. They saw an extreme landscape whose bizarre behavior required further explanation. According to the French *philosophe* Buffon, excessive, equatorial rainfall enervated people, plants, and animals (Gerbi 1973:14). Neither crops nor livestock would fare well. In the theory of the day, the land itself was still drying out, having only recently emerged from the ocean. The Andean cordillera provided further evidence of the Americas' youth. It rose jaggedly upward—far higher than European peaks (Ibid.:62). Such unweathered, sublime geology indicated poor prospects for settlement and other forms of domesticity. Australia, which broke next into European consciousness, presented the opposite problem. Nineteenth-century explorers and surveyors found the continent too *lacking* in jaggedness: its dry outback contained few mountains and fewer rivers or lakes. "Australia is . . . indescribable," writes Paul Carter, summarizing the disappointment: "In so far as its nature is undifferentiated, it does not have a distinct character" (Carter 1987:44). "Australians are still learning to see where it is that they live," writes another local critic (Seddon 1997:71). Africa has provoked a similar puzzlement. Much of the land mass sits on a plain of between 1,000 and 2,000 meters in elevation. Few mountains rise, and, far south of the glaciers' imprint, few depressions hold water. Sublime in its sheer monotony, Africa just extends. Visitors and settlers coined their own clichés: "miles and miles of bloody Africa" in the east and "great spaces washed with sun" to the south.[17] Like fragments of Britain—frozen at the moment of first separation—many white settlers could not innovate aesthetically.[18] A continuous, though attenuating, Wordsworthian taste has informed and shaped white expression in and about Zimbabwe.

Anglophone white writers have, therefore, devoted considerable attention to specific landscapes and the problems they pose. The plateau's flat, arid grassland, interspersed with bushes and short trees, recurs in settler fiction and memoir, disrupting Euro-African assimilation (Figure 1.1).[19] With distaste and fascination, Stockley (1911:319) describes "Africa's rolling leagues of bush and rocks and empty, rugged, burning land." At midcentury, Lessing writes semi-autobiographically:

> A white child opening its eyes curiously on a sun-suffused landscape, a gaunt and violent landscape, might be supposed to accept it as her own, to take the msasa trees and the thorn trees as familiars . . . [However] This child could not see a msasa tree, or the thorn, for what they were. Her books held tales of alien fairies, her rivers ran slow and peaceful, and she knew the shape of the leaves of an ash or an oak, the names of the little creatures that lived in English streams, when the words "the veld" meant strangeness, though she could remember nothing else . . . [I]t was the veld that seemed unreal; the sun was a foreign sun, and the wind spoke a strange language.[20]

Figure 1.1 Zimbabwe's savannah, photograph by author, 2007

More than 50 years later, the same alienation prevails. Lauren, sister of the Zimbabwe-born memoirist Wendy Kann, has died in Zambia. Kann looks for a gravesite on the farm:

> We wanted a beautiful place—a place that reflected Lauren's soul . . . We tilted our heads and squinted at each side, trying to see beyond the bush . . . I turned to one side and strode deep into the long grass looking for something, anything. I finally found a tall forked msasa tree with two smaller msasas and spindly acacia close by. If you ignored the scrub and thorns, together they made a circle, a canopy, something like a glade. (Kann 2006:33)

Can white souls rest in African soil? They will, Kann suggests, but only with the assistance of a vivid imagination and good hiking boots.

How, then, can one account for the comparatively effortless success of white spirits and white bodies in the western United States? There, in an unglaciated, southerly landscape quite different from the Wordsworthian ideal, Anglophones have established a sense of belonging so total as to be nearly beyond question. It was not always so. The first Anglophones found Southern California as bizarre as the Australian outback. Its canyons and arroyos evoked a "deep Mediterraneaneity"—understood by Spanish colonizers but unintelligible to children of the glaciers (M. Davis 1998:10–14). Englishmen could easily have failed to assimilate. But military and medical conditions gave them an advantage. European rhinoviruses decimated native peoples even before settlers and armies could finish the job with outright genocide and expulsion. In demographic terms, much of North America became a "neo-Europe" (Crosby 1986:2). On the cultural plane, whites' forgetfulness and inventiveness promoted the same transformation. John Muir and Ansel Adams, for instance, represented Yosemite Valley not as vacated, emptied land but as *simply* empty, virgin land (Cronon 1995; Solnit 1994:215–220). There, the elimination of people enabled a similar evacuation of environmental history and meaning. Settlers filled the void with concocted traditions of wilderness and environmental stewardship. Those investments paid hefty dividends. Conservation allowed midwestern environmental essayist Wes Jackson to, as he puts it, "become native to this place" (Jackson 1994). Thus, beyond the imprint of the glaciers, whites did assimilate to the land, assisted by the gun and the common cold. Particularly in the United States, ex-Europeans claimed the unmarked status, now the chief form of white privilege in public discourse (Wildman and Davis 2000:661). In Africa, could light-skinned people achieve the same belonging without those advantages?

EMBRACING AFRICAN LAND

Sidestepping the thorny "native problem" almost—but not entirely—cleared the path to Rhodesian belonging. The environmental route still held challenges: real thorns, aridity, bright light, and the lack of shoreline. To belong, whites would continually need to surmount their own proclivity toward Wordsworthian and Burkean mildness.[21] In common with British Romanticism, this tropical task centered on *seeing* the environment (Cosgrove 1984). Authors, photographers, and painters had to find ways of observing, interpreting, and portraying the savannah as homelike. In Australia, the first British travelers had already confessed the absence of a much-anticipated inland sea (Carter 1987:92). Similar circumstances forced Rhodesians and, later, white Zimbabweans to describe arid topography as comfortable and comforting. The odds have been against any such discursive success, as the most sensitive South African whites readily admitted. "How are we to read the African landscape," asks J. M. Coetzee. "Is it readable at all? Is it readable only through African eyes, writable only in an African language?"[22] On one level, Coetzee reiterates the problems of nomenclature Anglophones encountered elsewhere beyond the imprint of the glaciers. From Australia to California to Brazil, English explorers found their native tongue inadequate to the task of describing novel terrain (Raffles 2002:101). For South Africa, Antjie Krog's aptly entitled memoir *A Change of Tongue* captures this sense of deep foreignness. "[T]he landscape does not let itself be told," she laments at the edge of the Kalahari Desert. "I have no language for what I see" (Krog 2003:251). At a deeper level, Coetzee and Krog cast into doubt whites' entire project of belonging: they may *never* understand and assimilate to African landscapes. Nonetheless, white Zimbabweans—assisted by the writers and painters among them—have continued to muddle through. With considerable effort, they have crafted sensibilities for appreciating African land with European eyes and for narrating it in English.

Images often succeeded where words failed. Rhodesian painters chose nature as their primary subject and, through it, confronted the aesthetic challenge head-on. Robert Paul—who emigrated from England to Rhodesia in 1927—garnered praise for his early success. "He found form, cohesion, variety, vitality in that seeming nothingness," gushed one critic (Roux 1996:60). Nothingness was in the air, or so a fellow watercolorist opined. Writing of Paul's and his own technique, Martin van der Spuy summarized the problem of an empty, overly lit sky:

> The tradition of landscape painting originated in a moist climate. Atmospheric perspective is dependent on moisture in the air. In tropical countries the far distance is as strongly present as the foreground . . . [and so] the achievement of depth has to be tackled in a different way. (quoted in Kirkman 2007:26)

Van der Spuy does not disclose his own method, but it appeared to persuade the viewers of his art (Figure 1.2). "One can smell the dust, the air of the Zimbabwean landscape traced with scent of wild flowers," wrote a critic after van der Spuy's death (Bestall 2007:6). A final white artist located the problem of representation not in waterless space but in water-starved soil and vegetation. Jean Hahn emigrated to Rhodesia in the 1950s. "When I first came from Europe," she recalled in an interview in 1995, "I thought everything was terribly dry and dusty . . . and then I suddenly realized that that was Africa." She developed a palette of earth tones—tans, browns, and

Figure 1.2 "Goromonzi bush," painting by Martin van der Spuy, ca. 1998–99
Source: www.Zimbabwefineart.com.

reds. Indeed, so well did she assimilate to these hues that, on a return visit north, "the grass was deep green and I thought it was frightful!" (quoted in Murray 1995:10). Moderation be damned: savannah aesthetics had become home!

Another visual technology, the aerial perspective, contributed further to whites' sense of familiarity with and in Africa. Elevation allowed one, at least, to comprehend emptiness and vastness—and to come to value them. Here, words served better than images, as the heroine in Keith Meadows's adventure novel learns. Standing at the edge of the Zambezi escarpment, "staring out over the grey-green ocean of bush that stretched away into infinity . . . Peta had tried to capture the vista on film several times in the past, but had failed. It was just too big" (Meadows 1996:504). Such assertions of unrepresentability actually served to represent the bush. Also, if Africa cannot be captured in an image, then one must experience it. Elsewhere, Meadows suggests the exclusivity of such knowledge. In a set of biographies subtitled *White Africans in Black Africa*, he describes "a special fraternity that bonds pilots who frequently fly small planes over the spaces of Africa . . . an unsung brotherhood born of . . . compass bearings and unblemished wrap-around horizons" (Meadows 2000:90). This fraternity, Meadows does not need to say, hardly extended beyond white society: whites virtually monopolized Zimbabwe's pilot licenses, and a surprising number of whites owned light aircraft. Blacks could only guess at their vistas. Indeed, the aerial view opened up a parallel, but different, continent. "Nigel loved to set off [in his plane] over the land where for so long white men had fought and scrabbled to secure a foothold," writes Christina Lamb in her biography of a farming family:

> But his Africa was not the dark one of Dr. Livingstone. His was the Africa of the long horizon that seemed to stretch to the very ends of the earth. He flew over grasslands, salt lakes, swamps such as the Okavango Delta and Bangweulu Wetlands, the great rivers Limpopo, Zambezi and Congo, and the Kalahari Desert. *I felt privileged to call this home.*[23]

His was also the Africa—it would appear—that contained no blacks, no borders, and no nation-states. The "wilderness vision," to use Will Wolmer's term, skirted a blind spot for politics (2007:24, 145). Or, more precisely, it turned politics inside out, moving the white pilot from the margins of Zimbabwe to its center.

Over white-owned land, the bird's-eye view amplified this effect, making white wealth and power stand out. "[B]ougainvillea mark the houses of white men," writes Zimbabwe-born journalist Peter Godwin. "Bougainvillea is exotic to Africa, just like the white man. It hails from the rain forest of the Amazon. From the air, you can trace the progress of the European by the bright scarlet, mauves, and pinks of bougainvillea" (Godwin 2006:65). As a form of "imperial visibility" (Burnett 2000:126–129), this top-down perspective showed commercial production in stark relief. Farmers planned crops from air photos, regularly supplied by the Surveyor General (Whitlow 1988). Such images became icons (cf. Cosgrove 1988). Embellished with soil and crop notations, they graced the covers of agricultural manuals and hung in sitting rooms and farm offices. The bird's-eye view, then, carried the observer from mere comfort toward a truly proprietary ease on African landscapes. It converted minority status into majority ownership—and even seemed to blunt obvious challenges to property. Ian Holding's remarkable novel of the farm invasions traces the central family's Edenfields estate back to its visual beginning. At the time of the pioneers,

> old Oupa Jack . . . climbed to the top of the kopje [rocky hill] and was struck by the magnificent view out over the land . . . He had named the place Eden's View because he'd said to himself that he could plant as far as that view and no farther. The view was the boundary to the farm. (Holding 2005:169)

When occupiers have overrun Edenfields, the wife of Oupa Jack's descendent "believes there is some hope in this view, in this seamless stretch of sky, an aspiration that makes her not want to give in and flee to a foreign, unwelcoming place" (Ibid.:22). Death threats vanish in the haze, and the kopje makes belonging still seem tenable.

One final quality of the savannah, however, could upset this hard-won sense of home. As a material fact, the lack of standing water virtually offended whites: it was an environmental insult! To see water—not just vapor or vegetation—where there was none would require a heroic feat of imagination. Later in Ian Holding's narrative, Oupa Jack's great-grandson is wandering at night. The boy, Davey, is in crisis: thugs have killed his parents and taken over Edenfields. He intends to kill them. En route, "[T]he surroundings remain vacant, impenetrable constant . . . He is a boy walking across Africa." He is a *white* boy, it goes without saying. "So he compensates, formulating an imaginary landscape around him. He is walking on a sandy white beach. A blue sea is breaking over him . . ." (Holding 2005:146–147). In his mind, Davey has fled to a foreign—but welcoming—place. Aridity, then, goes hand in hand with armed aggressors (and here the landscape serves as both a metaphor and as its literal self). In a more peaceful context, Zimbabwe-born naturalist Michael Main pushes poetic license still further. His photo book on the Kalahari begins with geological data: the Magkagikgadi Pans are a paleo-lake that has dried out completely over the past 50,000 years. Today, only the largest floods deposit a thin sheet of water (Main 1987:18). This empirical recounting soon gives way to the romance of prehistory: on so-called Kubu Island (Figure 1.3), there is

> a pebble beach and it speaks of a time long, long ago when the pan was a sea . . . There were times when the whole island was deep under water and others when it barely showed, and there were times, like today, when it lay bare and naked to the sun, an island in a forgotten sea. (Ibid.:26)

The evocation plays with grammatical tense, conjuring into actuality a long-gone waterscape. Writing, then, could make up water where there actually was none—perhaps the most valuable technique in whites' imaginative project.

Savannah sensibilities were, thus, a work in progress. Even before paramilitary men swept them away, white farmers—in reality and

Figure 1.3 "Solitary sentinel in Kubu's shore," photograph by Michael Main, 1987
Source: Mike Main email: mmain@info.bw.

in fiction—continued to grapple with the unfamiliar. The American Wallace Stegner writes of even more arid lands beyond the Mississippi:

> A process of westernization of the perceptions has to happen before the West is beautiful to us. You have to get over the color green; you have to quit associating beauty with gardens and lawns; you have to get used to an inhuman scale; you have to understand geological time. (Stegner 1992:54)

White Zimbabweans did not quite make these adjustments. Art, fiction, and the entire imaginative project could not entirely replace the garden aesthetic still surviving from Britain. Whites depended too much on *seeing* the land, especially from high above it. Engineering would have to complete the job that representation had begun: whites dammed rivers and filled reservoirs for hydropower and irrigation. As a by-product, these industries incised Wordsworthian shorelines. And, as the rest of this book explains, whites rejoiced in the country's new waterscape. "You can bring any creature in the world to one of my dams, and when he sees that green grass and blue water he'll think it's beautiful," boasts the main character in *The Dam.* "*That's* what I call beauty" (Fisher 1973:42; emphasis in original). Through such conflations of aesthetics and infrastructure, biography and place coproduced each other (Pred 1986:21). Belonging ensued—a particularly geographical form of it. For a time, love of landscape allowed whites to root themselves in Zimbabwe and in other parts of Africa. Later, precisely that geographical obsession and its attendant social blindness exposed whites to tremendous political risk in the black-ruled nation.

* * *

Thus, whites and the writers among them translated inhospitable circumstances into home. In and of itself, this process is not unusual: immigrants assimilate and integrate everywhere. To do so, they negotiate their status from outsider to guest to peer. Colonial whites, however, could not stand the embarrassment of such pleading—and the attendant consciousness of their minority position. They dodged human qualities altogether and negotiated their status vis-à-vis the environment. "[T]he pull of wild nature," writes Keith Thomas, "can always be recognized as an essentially anti-social emotion" (Thomas 1983:268). Still, nature-loving Euro-Africans could have borrowed from environmental African meanings, such as the origin myth of "Guruuswa," or the long grass (Beach 1980:62–63). Perhaps too few whites respected Shona people and culture enough to support such syncretism. Instead, they sought to import European, particularly British, aesthetic sensibilities. The spirits of Burke and Wordsworth helped Anglophone settlers in their landscape-focused imaginative project, but only in the broadest sense. Regarding specific topographical and hydrological features, these traditions cast the savannah in a critical light. Never glaciated, the bush lacked water and varied terrain, and the sun beat down on it. Such deficits forced painters and writers to innovate, to develop their own "period eye . . . [or] culturally specific way of seeing" (Gaskill 1992:183; cf. Allina 1997:23–24). Making a purse from a sow's ear, they celebrated vastness and monotony. Whites climbed and flew, the better to situate themselves and their property within the savannah. When all else failed, they excavated archaic lakes or simply hallucinated water. Of course, these wet dreams did not make the veld any easier to farm and ranch. Desiccation plagued the settler's bank account even when it had ceased to trouble his eye. Eventually, however, aridity caused fewer and fewer whites to question their fit with the landscape. If they had once half-suspected themselves of trespassing, now they felt entitled to farm and to own farms.

That change of consciousness depended upon what might be called a political technology of belonging—one that functioned clumsily and only up to a point. The love of landscape was not "innocent." As Shiva Naipaul writes, Blixen's reminiscences of Kenyan beauty represented "an act of arrogation, an assertion of implicit overlordship" (Naipaul 1980: 147–148). Like a special optic, this aesthetic sensibility threw blacks out of focus while zooming in on landscape, plants, and animals. The process of selection worked: Euro-Africans overcame the limitations of minority status. They skipped over the triangular period—or Americanist "middle ground"—of negotiated meanings between immigrant and native. The attachment to landscape gave a white enclave the hubris of a white nation-state. But the trick did not fool blacks—or even white dissidents. "Africa is the Land of Wide Empty Spaces," lampoons Kenyan (black) writer Binyavanga Wainaina. In white fiction, "African characters should be colorful, exotic, larger than life—but empty inside . . . Animals, on the other hand, must be treated as well rounded complex characters" (Wainaina 2006). Francesca Marciano lodges the same criticism of white tastes in

Kenya: "Adventure, sex, beauty . . . is crap, interior decoration, advertising," one of her characters rants after the death of a friend in a road accident. "Because Africa is a fucking drunk Kikuyu in a Nissan" (Marciano 1999:164–165). Euro-Kenyans, Marciano implies, ignore blacks at their peril. She does not demand an end to racism. On the contrary, in her framework, even explicit racism—against the "fucking Kikuyu"—would represent an adaptive step. As long as they practiced escapism and disregarded the Other, ordinary whites could not even see the truck heading toward them. Extraordinary liberals did anticipate and adjust to both nationalism and antiwhite sentiment. With greater social curiosity—even if tempered with ethnic chauvinism—a larger portion of Euro-Africans might have been able to anticipate and weather these storms. Better still, with a more pluralistic form of social empathy, they just might have become African citizens in the fullest sense. In the event, by writing themselves so single-mindedly into the landscape, many whites wrote themselves out of the society.

It is a common, if underappreciated, ex-British predicament. "White settlers," argues Nigel Clark in a global comparison, "have been grappling with terrain that is so often resistant to European ideas and practices" (Clark 2005:365). Although philosophically primed for travel, Britons have frequently resisted the change of social or topographical scenery. In the forbidding deserts of western North America, Stegner and others marvel at the extent to which settlers have adapted aesthetically. Some environments differ less from the mother country. In mostly temperate New Zealand, white sheepherders have gone to court to demand the rights of indigenous people (Dominy 2001:207ff). Asserting their knowledge of and affinity for upland pastures, they have tried to occupy what Tania Li calls "the tribal slot" (Li 2000). Achieving varying degrees of success, then, such symbolic, discursive, and literary work has proceeded without plan or solidarity in the zones of European settlement. Emigrés scattered physically from the British Isles have reconverged around a structure of yearning. After movement, they seek emplacement (Orlove 1996). After cosmopolitanism, some seek autochthony (cf. Geschiere and Nyamnjoh 2000). That objective has proven elusive, but, on some frontiers, Europeans have at least established normal citizenship. Genocide helped naturalize whites as the archetypal North Americans. They belong in every habitat "from sea to shining sea." Indeed, as Bruce Braun argues, American popular culture continually encodes hiking, mountain climbing, and nature altogether as white

(Braun 2003; cf. M. M. Evans 2002). And few critics suggest—in the spirit of Shiva Naipaul—that nature first belonged to other peoples. In North America, then, ex-Britons succeeded in the enterprise still ongoing and possibly collapsing in Zimbabwe. White America realized the home-making, imaginative endeavor of white Africa.

PART 1

THE ZAMBEZI

CHAPTER 2

ENGINEERING AND ITS REDEMPTION

The savannah pushed whites' imagination to its limits and beyond. As the last chapter explained, many sought, in the bush, an escape from the intractable problem of minority status. Since they could not belong comfortably to African society, such whites sought to belong to African ecology. Yet, the landforms themselves seemed to repel whites' embrace. As children of the glaciers, Britons and other northern Europeans appreciated a well-watered, Wordsworthian topography. Much of Africa—especially Zimbabwe—offered a prospect diametrically opposed to this ideal: arid plains stretching unbroken in all directions. Of course, painters and writers such as Jean Hahn and Keith Meadows patched together a sensibility of the vast. Through it, they and at least some of their readers came to value "miles and miles of bloody Africa." But most whites, including conservationists, valued it only up to a point. Take, for example, Doris Lessing's novel *Landlocked* (1958b). The protagonist, a Communist and multiracialist of the 1940s, has many reasons to leave Rhodesia—including its hydrology:

> [Martha Quest] was becoming obsessed with the sea, which she had not seen, did not remember ... An enormous longing joy took possession of her. She no longer thought: I'm going to England soon; she thought: I'm going to the sea, I'm going to get off this high, dry place where my skin burns and I can never lose the feeling of tension and I shall sit by a long, grey sea and listen to the waves break ...[1]

Lessing herself left Rhodesia for London in 1949. What of those she left behind? As long as they lusted for seas and breaking waves, could true settlers bind themselves to the African savannah? Aridity still threatened to undo the white project of belonging.

Fortunately for whites, the colony disposed of tools beyond the merely literary: it could engineer water where there was none. Almost as soon as *Landlocked* went to press, construction firms made its title an anachronism. Between 1955 and 1959, ten thousand workers built a hydroelectric dam across the Zambezi, Africa's fourth-longest river, draining the continent's fourth-largest basin (see Figure 2.1).[2] In the next five years, water flooded 5,580 square km, creating what was then the largest reservoir in the world.[3] Concrete, in other words, did the job of ice sheets and gave Rhodesians an inland sea. Thereafter, the Central African Power Corporation[4] managed the reservoir. A formula known as the "rule curve" maintains electric generation by regulating flow through the turbines and over the

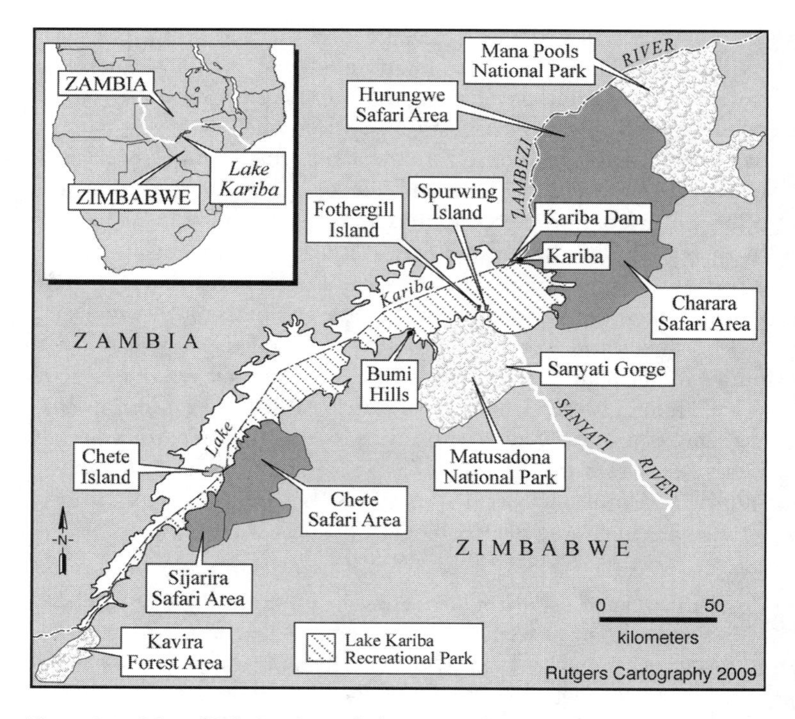

Figure 2.1 Map of Lake Kariba, including national parks and protected areas

Source: Map by Michael Siegel, Rutgers Geography Department.

spillway. This regime determines the water level and—as topography varies—the shape and length of the shoreline. And that shoreline is long, indeed: roughly 2,600 km. Of course, engineers blocked the Zambezi for economic rather than aesthetic reasons. Electricity fell within the administrative project, as an adjunct to black industrial labor. Firms involved in the dam advertised themselves as "making their contribution to a new and greater prosperity."[5] Less positively, the displacement of 57,000 Tonga speakers, flooded by the reservoir, also challenged whites in their effort to manage Africans. On the northern bank, Northern Rhodesia (and independent Zambia after 1964) dealt with the dam and Tonga strictly within this administrative framework. They enabled valley Tonga to become lake-fishers and businessmen (McGregor 2009:112). Such straightforward political economy did not obtain on the southern bank. There, where whites were more numerous and clung to power longer, the imaginative project disrupted and complicated matters of fish and finance. Conservationists claimed the foreshore and cordoned it off from almost all development, much to the detriment of local Tonga. Ever since, they have struggled to farm drier soils significantly inland from the lake. For many whites, on the other hand, Lake Kariba fulfilled dreams of the "long, grey sea" lacking in Lessing's Rhodesia.

But it did so only after considerable effort and delay: at first, Kariba Dam represented an assault on the very values undergirding whites' emergent identity. How could conservationists and lovers of African landscape embrace the industrial impoundment of the Zambezi? After the wall's closure, rising water killed all but a fraction of the animals and drowned all plant life. In 1959, Reay Smithers, director of the National Museums of Southern Rhodesia, decried Kariba as "the greatest environmental upset ever to befall a population of animals and birds within the African continent, in the memory of man" (Smithers 1959:21). In the 1960s, writers—all of them white—documented the ecological destruction. (They hardly noted the immiserated Tonga.) Gradually, however, another agenda came into view. Another generation of authors—still all white and almost all male—reimagined the lake. Beginning in the 1970s, they wrote movingly of the "unspoilt Africa" and superb angling found at Kariba.[6] To the extent that these works acknowledge the dam at all, they paint a picture of nature restored and enjoyed. Euro-Africans yearned for water, glorified the lake, and forgave the dam. Literary and photographic arts now polished Kariba's rough edges. Aesthetics redeemed Man's industrial sin against the Zambezi River, reconciling the abundant contradictions. White settlers remade the savannah in their image, and meanwhile loved it for being African. Because they loved it, they felt themselves to be Africans.

History, Geology, Nature

Writers on Kariba face the perennial problem of historians: how to set the lake and the dam in timelines of change and continuity. Do these aquatic objects represent a geological process, akin to Braudel's (1980) *longue durée*, or do they constitute events—*conjuctures*? By and large, geology, climate, and biology evolve ever so slowly, literally at a glacial pace. Such systems seem natural precisely because they move so out of step with human activities. They inhabit John McPhee's (1980:20) "deep time" of the U.S. West: the Pleistocene, Jurassic, Triassic, and all periods before humanity. Colonial authors tended to treat Africa in precisely this fashion—as what Anne McClintock calls "anachronistic space . . . exist[ing] in a permanently anterior time" (McClintock 1995:30; cf. Hammond and Jablow 1970:176). At Kariba, however, the shallow time of human history changed geology. Built in only three years, the dam stopped the flow of a river roughly two million years old (Main 1990:5). The reservoir bent the crust of the earth and—according to some—turned waterlogged trees to stone. The first generation of Kariba writers grappled with the implications of this merged, geohistorical time line. They used metaphors of domestication: Some represented engineers and workers taming the savage Zambezi River and Africa in general. They also used metaphors of degradation, appalled that, in such a short time, people could destroy what Nature had taken so long to create. This narrative anticipated more recent criticism regarding anthropogenic climate change (McKibben 1989). But the sense of regret did not last. Within ten years of the dam's construction, Rhodesian writers were cautiously associating the dam with nature, even situating it in a geological time line. Taken as a whole, the Kariba literature moved the Zambezi from geology to history and—in the course of absolving the dam—from history back to geology.

The Zambezi entered Anglophone literature in the course of an earlier, unsuccessful project of domestication: the effort to navigate the river and "open" Africa for evangelism, trade, colonization, and resource extraction. In 1860, while descending the river, David Livingstone encountered a "country too rough for culture." Rapids broke the smooth course of the Zambezi. "Some rocks in the water . . . dislocated, bent, and even twisted to a remarkable degree, at once attest some tremendous upheaving and convulsive action of nature . . . We have probably nothing equal to it in the present quiet operations of nature."[7] Local residents, Livingstone related, called the place "Kariba." Subsequent visitors learned that the term, in fact, referred to a rock beneath the rapids, where the river god Nyaminyami was said to live. Shaped like a snake, Nyaminyami arched his back and otherwise agitated the waters in fits of emotion. Livingstone and later explorers treated this animism as quaint while they focused, in deadly seriousness, on navigational issues. Kariba entered colonial geography—or in Carter's (1987) terms, became a feature—because it accelerated water beyond the capacity of European haulage. Indeed, the first published photos of the river—taken on a 1903 voyage downstream—depicted cataracts and the gorges that flanked them. Eighteen of 26 images and captions in de Lassoe's report to the Rhodesia Scientific Association showed or referred to Kariba, Gazongo, and the Kebrabassa cataracts (de Lassoe 1908). At roughly the same time, another explorer, who had surmounted Kariba going upstream, recommended the construction of a lock and dam at Kebrabassa. Thus, both gorges gained recognition as sites of wild, unnavigable water.

The second wave of writing on the Zambezi—a good half century later—also sets wilderness against technology, but with much richer meanings. As plans for the dam advanced, these works struggled to place Kariba in a time line, first geological, then historical. Written in 1954, *Crocodile Fever* relates the "true story" of the South African–born hunter Brian Herbert Dempster, as told to Lawrence Earl in London.[8] In 1947, Dempster and two African assistants ascended the Zambezi to a pool in the Kariba Gorge. Like Livingstone before him, this European associates the landscape with remote antiquity. "It was as if even nature were standing breathless before this prehistoric scene," writes Earl. "Dempster, held in a kind of awed homage, felt he was intruding on a past millennium" (Earl 1954:97). The hunter shakes off his awe forthwith and shoots and skins crocodiles for a good six months. As Dempster dispatches the last one, Earl (1954:141) reflects:

> Later—after Dempster had left the district—surveyors for the great electric power project would come, dispelling the loneliness of the gorge still more. Probably never again would the saurian giants make of the shadowy divide a forgotten retreat into a prehistoric age.

The passage places Kariba squarely in the era of dinosaurs but, meanwhile, on the cusp of a violent temporal shift. If Time had "forgotten" the gorge, the dam would soon remind Time of it. Earl, thus, anticipates a feeling of pastoral nostalgia (cf. R. Williams 1973). He sketches a preindustrial—in this case, prehuman—refuge and foretells its demise.

As that lengthy process unfolded, between 1959 and 1963, six journalists—four British, one American, and one Rhodesian—visited Kariba and published monographs in London and New York.[9] Their accounts report actual events but, through the use of metaphor, they manipulate time lines in imaginative ways. The American, David Howarth, explicitly signals the end of geological, deep time. In the first half of the twentieth century, "time in the Zambesi Valley was almost standing still." However, 1955 "might almost be said to be the date when history began. It was then that the final decision was taken that was to bring the existence of the valley to an end" (Howarth 1961:1, 29). The Zambezi did not die easily. It defied its own geological timescale by flooding to the 1,000-year level in 1957 and to the 10,000-year level in 1958. These surges imperiled men, machinery, and the dam site itself, jeopardizing the entire project. The 1959–63 published accounts personify the river—or deify it, suggesting that Nyaminyami is fighting for its freedom. Geology is battling against history, deep time against shallow time. Frank Clements, the single Rhodesian author, codes these forces as black and white. The French engineers and Italian construction foremen

> have been matched against a force which, while it seemed blind and barbarous, was also magnificent, and which they discover they have also learnt to admire. Although the river has been blocked and the great bastion at Kariba stands as a monument to the white man's genius, there are few in Africa who would claim that Nyaminyami has been defeated, and there are many who believe he will yet have his revenge. This curious conflict in the souls and minds of men accompanies on another plane the physical struggle to master a continent, a great part of which is still a survival from the world of pre-history.[10]

Yet again, colonial expertise brings civilization to benighted Africa.

Such triumphalist narratives of Progress render the Tonga invisible. Their elaborate system of floodplain agriculture—well described by Elizabeth Colson (1971) and Thayer Scudder (1962)—predated the dam, and the dam destroyed it along with the Gwembe Valley. Elizabeth Balneaves, the only female writer of 1959–63 and the only one based on the northern bank, writes with unusual sympathy for the Tonga. Centered on one of the participants in Operation Noah, her account includes a photograph of water inundating a Tonga village.[11] With one exception, all the 1959–63 works mention the signal act of Tonga resistance: refusing to move from Chief Chipepo's area, on the northern bank. In an armed riot, colonial police killed eight Tonga, a tragedy the authors blame on the agitation of African nationalists from Lusaka. Presumably, the loss of one's home and livelihood could not, in itself, motivate rebellion. According to Colson, forced resettlement disrupted Gwembe society in every sense. In Zambia, these displaced people have recovered to some extent. Many have joined the fishing industry at Kariba's shore. Artisinal fishing camps dot the Zambia bank and roughly 50 percent of the Zimbabwean shoreline. The other half of the southern littoral falls under protected areas (Figure 2.1). Indeed, Zimbabwe's Lake Kariba National Recreation Area prohibits cropping and stock-raising along the entire shoreline up to 3 km inland. Tonga, therefore, mostly live inland, where they cultivate some of the driest soil in Zimbabwe.[12] Now or then, few Tonga would agree with Clements's crass triumphalism. "Here," he declares, "in what was a savage wilderness, man has come to stay . . . " (Clements 1959:199). "Man" means whites, and whites mean modernity.

Yet, history does not end so quickly. The closure of the dam wall in December 1958 generated a final flood and more complex reflections on time. Watching the river rise, authors made comparisons with the Old Testament. As a metaphor, biblical time mediated between extremes of geology and history. Charles Lagus finds a tortoise floating on the growing lake: "it gave me a sad, reproachful look from its wrinkled antediluvian eyes as though it had escaped the First Flood only to come to this—a deluge created by me and my kind" (Lagus 1960:103). Other animals fared less well. It was "something no naturalist had ever seen before," recalls *Animal Dunkirk*, signaling the rupture with geology. Before the "hungry maw of the lake," birds hatched eggs, watched the chicks drown, and renested on higher branches of drowning trees (Robins and Legge 1959:153–154). Fortunately, people could and did intervene—in a suitably biblical fashion. In 1959, the Game Departments of Southern and Northern Rhodesia launched Operation Noah and ultimately saved

7,000 animals.[13] White leaders of the Southern Rhodesian rescue mission feature prominently in the texts and photos of 1959–63. In an iconic image, the operation's leader, Rupert Fothergill, cradles an impala fawn against his bare chest.[14] His sanctification as "Noah" probably contributed to the outpouring of public donations toward the rescue and to the eventual establishment of the Department of National Parks and Wildlife Management, with Fothergill as its first director.[15] Kariba's flood *made* conservation.

Unlike the Flood, Kariba's waters did not subside, but they did stabilize. In literature, the valley returns to geological time and creates conditions for ecotourism. All of the 1959–63 monographs, except the more critical work by Balneaves, reach a moral and ecological resolution.[16] Robins and Legge continue their biblical metaphor, quoting Genesis: "I do set my bow in the cloud, and it shall be for a token of the covenant between me and the earth."[17] Since the waters remain, they create a lacustrine promised land—"a patchwork of plenty stretching to the horizon and born of the lake" (Ibid.:182). Kariba, the authors predict, will power industry throughout the Rhodesias, light every home, irrigate vast acreages, and support a rich fishing ground.[18] Such modernity usually incurs an environmental cost, but Kariba's aesthetics seem to violate this law. In print, the lake reconciles ecology and industry, embodying a spirit comparable to the "technological sublime" of San Francisco's Golden Gate Bridge (Nye 1994). Even the dam's curve mirrors a suspension arch, and both complement the scenery around them. Kariba, predict Robins and Legge, will inevitably become "one of the world's *natural* pleasure grounds for tourists" (Robins and Legge 1959:175; emphasis added). In order to accommodate the 7,000 mammalian relocatees, Southern Rhodesia designated the Matusadona National Park and Chete Safari Area along the littoral. Eventually, the Parks Department designated the entire shoreline as a recreational park (Figure 2.1). Recreation and tourism serve as a code word for the geological: "the lake," Robins and Legge write, "will eclipse the international tourist attractions of Kilimanjaro, Victoria Falls, and the Pyramids. By then, 'Operation Noah' will have passed into Africa's history" (Ibid.:183). Sedimented into the continent's past of deep time, the reservoir will take its rightful place in a rocky pantheon.

Perhaps the lake has always belonged to deep time. Two of the 1959–63 books suggest such continuity by citing a geological fable. As Clements writes:

> There are some who say that all man has done is to restore the ancient geography of Africa; that tens of thousands of years ago when the Zambesi ran westwards to the Atlantic, there was a lake which covered the valley between the hills of the northern and southern escarpment. (Clements 1959:213)

In truth, the Zambezi never emptied into the Atlantic and never before filled a lake.[19] Still, for those who believe it, the tale provides comfort. It allows one to consider the construction of the dam and Operation Noah as brief historical-biblical interludes in the otherwise stately geological procession. The tale also provides some moral justification for the death and destruction upstream of the dam wall. As Robins and Legge relate:

> Scientists now believe that Lake Kariba occupies the site of an earlier lake. Pebbles washed up on the shore are rounded as though washed by waves in past centuries . . . There was certainly an earlier exodus of animals, which without man's intervention, undoubtedly suffered a larger proportion of casualties than in the present flood. (Robins and Legge 1959:152)

Rather than obliterating the past, Man reenacted—with greater mercy—what cruel Nature had accomplished long ago. Such geological myths persist. In 2003 I met Eddie Daniels, who, as the chief topographical officer to the Surveyor General of Zimbabwe in the 1980s, had remapped the entire lake bottom with metric contours. Taking exception to my notion of environmental catastrophe, he corrected me: "It was once a huge inland lake anyway."[20] He had examined the polished pebbles himself. Thus, popular science and popular writing recast the history of environmental ruin as a morality play of ecological restoration.

FROM ACCEPTANCE TO ABSOLUTION

With the end of Operation Noah, foreign and local journalists left the lake to those with greater time to contemplate. From the early 1960s to roughly 1980, a series of ecologically minded whites grappled with concepts of progress, ruin, and restoration. Unlike the cohort of 1959–63, these writers did not reach tidy conclusions. More was at stake for them. In 1963, whites voted the conservative Rhodesian

Front party into power, bringing to an end the reformist "winds of change" period. The state ceased to consider any black involvement in politics. Then, in 1965, Rhodesia unilaterally declared independence from Britain. Thereupon, Britain isolated the country economically. By 1973, two nationalist guerrilla movements had begun to attack targets linked to the state and commercial agriculture. These developments encouraged whites to stick together in support of Rhodesia and things Rhodesian—perhaps including Kariba dam. At the same time, Lake Kariba became part of whites' lives in multiple ways. Located on the border with Zambia, the new town of Kariba served as a military base. A sizeable portion of Rhodesia's white male population passed through it, meanwhile admiring the lake. Hostilities did not prevent the onset of tourism, an industry that grew prodigiously after independence in 1980. Ultimately, the "tourist gaze" (Urry 1990) redeemed the dam, rechristening it as "wild Africa." Kariba's photo-literary archive promoted and reflected this process: a gradual, often contested, acceptance and naturalization of the reservoir.

More than any other writer, John Gordon Davis brought the lake into Euro-African daily life. Born in Rhodesia in 1936, Davis wrote his first book, *Hold My Hand I'm Dying* (1967), between 1962 and 1966.[21] In that period, the lake reached full capacity and Rhodesia divorced from the metropole and suffered her first attack by nationalist insurgents. *Hold My Hand* discusses all of these events, and its genre of macho adventure novels rooted in the tropics appealed to European and American readers. "This is the best novel coming out of Africa that I have read for a number of years," commented the prolific South African novelist Stuart Cloete, "It is Africa today. The characters develop in the skies and spaces of the continent. Love, battle, boredom, drink, all woven into the tapestry of Rhodesia" (J. G. Davis 1967:frontispiece). In short, the doings of whites—especially when at the lakeshore—acquire a new drama and importance.

Opening in the Zambezi Valley, the novel first describes Kariba with pastoral nostalgia. The protagonist Mahoney, serving as native commissioner and magistrate, has warned the Tonga of "a flood that would stay forever and drown the whole valley" (Ibid.:1). (Meanwhile, in full-throttle masculinist fantasy, a Tonga virgin submits coyly to rape by an Ndebele man.) Later, after trying the rape case, alcohol brings forth Mahoney's emotion. He loves "the bush" and his primitive subjects. He does not want them civilized and doubts the notion—of "this landlocked country"—that there should be a "partnership" between the races.[22]

> Africa, my Africa, is dying, like that Zambezi Valley down there, that mighty magnificent violent valley . . . It's going to be drowned by Progress. By Partnership . . . There's going to be no more sunset silhouettes as the animals come down to the mighty river to drink . . . There's only going to be the screams of the animals dying. There's going to be no more river god for them [the Tonga], no more Nyaminyami . . . There will be no more Batonka [Tonga]. They will just become bewildered Rhodesians. That's why it's sad, why Progress is sad. That's why Africa is dying, because the same sort of thing is going to happen everywhere. (Ibid.:30–32)

In subsequent chapters, the narrator sketches the construction of the dam, the floods of Nyaminyami, and the resettlement of the Tonga. Mahoney joins Operation Noah, and Davis presents a more gruesome view of the destruction than anything written in 1959–63. Mahoney and his trusted African sidekick come to a half-submerged tree, on whose branches hundreds of starving monkeys are clinging, water rotting the flesh of the lowest ones. Mahoney cannot rescue them as they instinctively attempt to bite him. So he puts them out their misery with a .22-caliber rifle:

> The carnage . . . Monkeys were blown to bits out of the trees, blood and fur and bodies flying, monkeys screamed and jumped and fell wounded into the water . . . Monkeys clung wounded to the naked branches and ripped open their wounds with their fingers and pulled out their insides . . . Again and again Mahoney fired through the trees, no longer sickened, only frantic to destroy and end the fear and the carnage. (Ibid.:94)

Yet, Davis's tone soon changes. Suzie, who becomes Mahoney's long-suffering girlfriend, flies to Kariba as one of its first holidaymakers. Descending to the lakeshore, she immediately appreciates its aesthetic qualities, all the more striking in the tawny, dry bush: "Blue, the water of the great lake was blue, like the sea and it seemed as big as the sea" (Ibid.:111). During her stay, she meets Mahoney and discusses the lake with him. "It's beautiful," she says. "Yes, and sad," responds Mahoney (Ibid.:127).

When I interviewed Davis—appropriately while driving along Spain's Costa del Sol—he recalled the evolution of his thought in the mid-1960s: "There was sort of an invasion of natural Africa, which I found sad, but I got used to that idea ... When the original dam was being constructed ... you felt it was a rape and an intrusion. Years later ... I felt it was a *fait accompli*, and [except from the Tonga perspective] a lovely *fait accompli*."[23] At the end of *Hold My Hand*, nationalist guerrillas threaten to undo that fact of engineering. The setting unites all the elements of Mahoney's personal and political dramas. It is the "night of the long knives"—the ever-anticipated moment when black servants *en masse* kill their white masters. Guerrillas have laid explosives on the dam wall, and Jake Jefferson—Suzie's husband and Mahoney's nemesis—must defuse them. Mahoney, himself, is battling insurgents on the Zambian border, downstream of the dam, where Suzie arrives to deliver their baby and die in childbirth. Amid this desperation, the dam assumes a positive, indispensable value.[24] Mahoney raves to himself:

> " ... the dam wall! If the black bastards blew that precious wall—Jesus, good Jesus Christ don't let that wall go up, not that wall, God!—the destruction, the unholy havoc, the biggest tidal wave in the world ... the holocaust". (Ibid.:504–505)

Jefferson saves the dam and proceeds, unknowingly, to raise Mahoney's son. Twice over, he rescues and perpetuates white-given civilization in Africa.

He also rescues a kind of nature increasingly associated with Kariba. By the 1970s, few could deny the beauty of the littoral. Even authors who pined for the wild valley betrayed a sense of wonder. In 1974 and 1975, U. G. de Woronin published a series of articles in Salisbury's newspaper *The Sunday Mail* and republished them in one volume in 1976. Born to the Francophone aristocracy of tsarist Russia, de Woronin had fled the revolution as a boy and, via England, emigrated to Rhodesia. Throughout the 1930s, he had hunted in the Zambezi Valley, sometimes with Davis's father. On one level, his published recollections of this time express a straightforward nostalgia. Entering a particularly remote area, the senior Davis encounters geological time: "if a dinosaur walked out it would not surprise me," he says to de Woronin (de Woronin 1976:22). Four decades later, de Woronin misses the ecological abundance of the free-flowing Zambezi: "Before [the dam], raging floods inundated many hectares of river bank, depositing rich silt, which grew lush grass ... " Now, the controlled river nourishes only short, sparse grass. Animals lack sufficient browse. Elephants, in particular, must uproot grass and chew through the adhering dirt. "Their teeth can be heard squeaking on grit from far away," as they wear down seven sets of molars toward premature starvation (de Woronin 1976:11, 17). Despite this bleakness, de Woronin acknowledges Kariba's surpassing beauty. He entitled his series in *The Sunday Mail*, "Anecdotes from Aquarelle." As explained in the forward to the volume, de Woronin and his wife christened their house with this name, French for "watercolor," because it "overlooks a vast expanse of the Lake and its ever-changing hues"—literally, water colors (Ibid.:2). In other words, each memoir of the valley first compliments the seascape that destroyed it. Kariba defied easy judgments.

In 1978, the biologist Dale Kenmuir intervened in popular debates to clarify matters. His *A Wilderness Called Kariba*—the most scientific study to reach the broader public—sold 6,000 copies in three printings and is still widely read.[25] From his base at the Lake Kariba Fisheries Research Institute, Kenmuir observed a biological recrudescence: "sponges, shrimps, sardines, mussels, clams, jellyfish, gulls, terns and turtles," he writes, "makes [*sic*] the phrase 'Zimbabwe's inland sea' much nearer to the mark than most people suspect!" (Kenmuir 1978:107). As if this were not enough to celebrate, the lake still retained pre-dam features. Elephants followed old paths, swimming above where they had once walked. Eels, Kenmuir reports, have managed to migrate upstream, through drains in the dam wall (Ibid.:94–95). Hydropower can even enhance biodiversity: Although, "a sudden discharge of water from the flood gates kills fish in the stilling pool, . . . further down-stream these waters are used for life-giving spawning purposes" (Ibid.:139). Kenmuir concludes *A Wilderness* with a plea for regulated, seasonal discharges. Yet, in a later work, of children's fiction, the hero times a discharge opportunistically to catch a trans-Zambezi rhino poacher (Kenmuir 1991). At his home—outside Cape Town, overlooking the ocean—Kenmuir described this manipulation of the river for crime fighting as a bit of "poetic license."[26] Perhaps even more poetically, he urged me, "We shouldn't think of it [the reservoir] as artificial. The only thing that is artificial about it is the dam wall." And, he continued, the obstruction could have been caused by natural means as well—although, he admitted, the floodgates could not have been.[27] In this partial sense, the dam only reproduces on the Zambezi what geology has accomplished elsewhere. Kenmuir's science—sometimes supplemented by speculation—brings hydropower one step closer to ecological respectability.

With a more mystical language, another publication of 1978 further redeems the dam. Born in South Africa, Alf Wannenburgh worked as a newspaper journalist, until—in a "hippie phase" of his life—Struik Press asked him to contribute text to a photo book entitled *Rhodesian Legacy*. Just down the beach from Kenmuir's house, he explained his own theory of naturally arising barrages: "the dam could just as easily have occurred through an earthquake [and landslide] that blocked the valley . . . Man acted as a force of nature."[28] This counterfactual conjecture fit the botanical—rather than geological—data rather more closely. Whereas, in the 1960s, the rule curve's regime of dramatic drawdowns had killed vegetation along the lake, in the 1970s, a swampgrass—*Panicum repens* from the upper Zambezi—colonized the shoreline. By 1977, when Wannenburgh visited Kariba, the once inhospitable "soak zone" teemed with buffalo, other ungulates, and their predators.[29] This biodiversity provoked an excitement that Wannenburgh could scarcely contain in print: "witness and take part in . . . the evolution of an entirely new African wildlife ecology," he exhorts the potential visitor.[30] Was that ecology truly wild, I asked the author? "What has been created [at Kariba] is something that nature can develop," he explained.[31] If that is so, Man has played the role of God in the Enlightenment's "divine watchmaker" thesis. Once created, ecology took its own course. In *Rhodesian Legacy*, Wannenburgh hints at this cosmology indirectly. "Now," he reports with a markedly unjournalistic reliance on hearsay,

> some Tonga say there are two Nyaminyamis—one for the deep water and one for the shallows . . . When the waters are blue and unruffled . . . they say that Nyaminyami is contented in his situation. But when the sky is black with clouds, and the lake is lusterless and opaque . . . they say he regrets having agreed when he did to live out of reach of the sunlight.[32]

If Nyaminyami symbolizes the will of African Nature, then, on a good day, Nature appreciates the full package of concrete, reservoir, and *Panicum*. Indeed, as Wannenburgh slyly insinuates, Nyaminyami accepted the idea from the beginning.

Two years later, the novelist Richard Rayner issues an even more magnanimous pardon in Nyaminyami's name. *The Valley of Tantalika* (1980)[33] concerns a river otter and a herd of impala imperiled by the rising waters of the reservoir. Narrated mostly from the animals' point of view, the story follows the river otter, Tantalika, as he seeks an explanation for the flooding from his deity and tries to understand white men. He consults Fura-Uswa, the animals' equivalent of Nyaminyami, who tells Tantalika that humans "are trying to destroy all living things in the Great Valley and even the valley itself."[34] Tantalika warns the impala to seek higher ground. Yet, as the flood begins, the narrator shifts perspective. Rayner writes, "There was a strange and terrible beauty about the spectacle of a valley which had lived, vibrantly, for millions of years, slowly drowning by the hand of man, to serve him in its death" (Ibid.:74). Soon, humans—explicitly "pale" men—reveal further unexpected capacities. They assist Tantalika in his rescue mission. A human, surely modeled on Rupert Fothergill, saves Tantalika himself from drowning. Fura-Uswa is "almost pleased with what they were doing," relates Tantalika (Ibid.:162), and Rayner's epilogue completes the moral resolution:

> The Great Valley is at peace now; the deep wound which man inflicted is healing, for he has come to terms with Nature and Nature, perhaps, with him. Both are receiving the benefits from the forming of the lake. For man there is the hydro-electric power for his homes and industries in Zimbabwe and in Zambia, with a vast playground for his leisure, where he can gain new knowledge of Nature's ways. For Nature, there is a sanctuary for all her children of the Great Valley in the areas which border the lake [Matusadona and other protected areas], and there is the lake itself. (Ibid:164–165)

Wild animals redeem the lakeshore, and the lake redeems the dam.

Is this resolution too neat? Dick Pitman, mentioned in Chapter 1 and the author of the last of the 1978–80 works, admits as much and confesses to self-deception. *Wild Places of Zimbabwe*—a meditation on nature and tourism that sold 10,000 copies—devotes a chapter to Matusadona National Park. Pitman, who had emigrated from Britain in 1977, labels Kariba unflinchingly as "one of the most massive ecological disturbances yet created by mankind." "However," he proceeds, "nature has made a spirited attempt at putting matters right"—largely through the *Panicum* grass—and "much of the Matusadonha shore has a unique and haunting beauty" (Pitman 1980:164; cf 1983a:10). In Harare, I probed these thoughts with Pitman. "The attraction is very largely aesthetic," he admitted. "It has very little to do with functioning ecosystems and all that. Looking behind the veil . . . [the lake is artificial] . . . but you get there and sit there and it doesn't matter. I mean how much reality can you handle?"[35] His candor startled me, especially given that in 1982 Pitman founded the Zambezi Society precisely to fight another dam. He and the organization's 90-percent-white membership contributed to the defeat of proposals for the Mupata Dam downstream from Kariba.[36] Why would he support one dam and not the other?[37] For Pitman, something changed between the closing of Kariba's dam wall and his publication of *Wild Places*. He ends the book by arguing that urban Southern Africans—white and black—increasingly appreciate wilderness, or places that appear to be wild. Like John Muir in the industrializing United States, Pitman sees such locales as an antidote to technological life.[38] He foretells, "the appreciation of beauty will grow . . . [and] the need for refreshment seems certain to grow accordingly" (Pitman 1980:190). Pitman was right: after the war, tourism boomed at Kariba.

But why should Kariba, in particular, in comparison with other venues, refresh whites so thoroughly? Certainly, it went a long way toward correcting Zimbabwe's aquatic deficit. Although still lacking in natural lakes, the country now boasted an artificial lake on a par with any in the world. Indeed, Kariba exceeded natural lakes in the intricacy of its littoral. Among the ecological benefits Pitman cites, were "additional habitats offered by the *indented* Kariba shoreline" (Pitman 1980:176; emphasis added). He continues, "a host of bays and inlets, can swallow a hundred boats without a trace" (Ibid.:178). I pursued this question of aesthetics and recreation with Rex Taylor, the lake's leading nautical authority. Since 1992, he had written the "Kariba" page in *Zimbabwe Fisherman*, an angling magazine for an almost exclusively white audience. In 1997, he had recommended a particular sailboat whose "shallow draft allows the captain a far greater choice of mooring sites in hidden bays" (R. Taylor 1997:26). "Why hide," I asked? "Privacy," he responded, "because that is what you go out on the lake for."[39] Presumably Taylor's readers shared this retiring tendency. In 1999, he wrote of "a fantastically secluded bay, cut off from all sight and fury of the waves! It was an incredible find" (R. Taylor 1999:40). At his home, overlooking the lake, Taylor showed me a sketch map of his favorite spot. There, Kariba's water extended far inland, rounding a corner, and ending in what he described as a closed "amphitheater." Like Pitman, he associated such isolation and short sightlines with wilderness—perhaps with the waterholes that had already become central to wildlife photography (Bunn 2003). In fact, the relationship was illusory. Young by definition, most reservoirs incise jagged, dendritic lines—an abundance of the boundary of land and water that Wordsworth so appreciated. Over geological time, erosion will round Kariba's corners, shaping it into an enormous waterhole. In the short term, then, Kariba displays what one might call a *geometry of beauty*, but one that still is patently artificial. Pitman, Taylor, and other whites conflated engineering with its exact opposite, nature.

From that conflation, however, ensued an even more meaningful engagement with the lake. Whites felt they belonged there. Margaret Peach entitles her 2003 memoir—centering on Kariba in the 1960s—*My Place in the Sun*. She writes, "The sparkling blue water against the hazy backdrop of the Matusadona escarpment gives one a lovely 'welcome home' feeling" (Peach 2003:5). Peach, who now lives outside Cape Town, declined to meet with me (the only Kariba writer to do so). In an email, however, she explained that quotation. "The saying goes," she recalled, " 'If you haven't been to Kariba you haven't lived.' I suppose it's rather parochial but then Kariba is [to] the Zimbabweans what the seaside is to other folk who do not live in a landlocked country."[40] Such sentiments in fact stood parochialism on its head: lovers of Kariba imported to Africa's interior the sensibilities of a continent's edge. In the wake of concrete, writers had manufactured a cultural home.

DISSENT, DISPLACED AND MUTED

Not all white authors accepted Lake Kariba or forgave the engineers. Three works written after 1990 mention the dam's disruption of the flood cycle downstream. The rule curve regulated flow to such an extent that the Zambezi hardly broke its banks. Especially in Mana Pools National Park, the floodplain of fresh alluvium shrank to a narrow band. Documented in scientific papers—and experienced personally by many Zambezi enthusiasts—these changes appear in the photo-literary archive.[41] Yet, the authors seem to pull away from drawing the obvious conclusion that Kariba destroyed a wilderness. Keith Meadows, for instance, sets his masculine adventure novel in Nemana National Park, transparently representing Mana Pools. The main protagonist, Harry Kenyon, has hunted elephant, rescued animals in Operation Noah, and turned park warden at Nemana. "Old haunts," reminisces Harry, "the Zambezi, scarred now by the cancer of the great wall that had spread resolutely across Kariba Gorge... Hundreds of miles of wild Africa gone forever" (Meadows 1996:173–174). That passage refers to points westward and upstream of the dam wall. Elsewhere, Meadows describes the damage eastward, downstream into Mana Pools: "There had been no floods since the wall blocked the gorge" (Ibid.:129). Bizarrely, this keen awareness of engineering and damage does not seem to affect Harry's sense of place. "Nemana National Park," Meadows writes in a stage-setting passage, "This was his parish. Almost 900 square miles of Zambezi Valley unspoiled wilderness. From the high hot barren escarpment... down to the Garden of Eden flood plains..." (Ibid.:57). When I met Meadows—in Bulawayo where he ran a safari business—our conversation reiterated much of the ambivalence running through *Sand in the Wind.* Meadows described Kariba as a "desert of water." Yet, he confined this indictment to the lake itself. In his mind, Mana Pools—after which he named his daughter—remained hydrologically inviolate. Except for the crowding of tourists and cars: the hospitality industry, rather than the dam, had reduced the park to "a rape status victim."[42]

If Meadows thus displaced his sense of grievance, other critics stifled their feelings altogether.[43] *Life and Death of a Pool* (1993), John Struthers's lovingly detailed photo essay on Mana Pools, makes no mention of Kariba or its effects. Born in Zimbabwe, Struthers moved to Stellenbosch, South Africa, where we met almost by chance, outside a bookshop. I asked Struthers whether he had ever published photos of Kariba. No, he said, but he had visited Matusadona National Park. There, he observed buffalos eating the lakeshore *Panicum.* This scene—the source of Wannenburgh's uncontained excitement—only saddened Struthers. "[B]ecause it is an unnatural situation," he explained. "They all got river fluke." Kariba's still waters harbored parasites normally killed in running water. In the animals' stretched hides, Struthers saw the monstrosity of an impoundment across the Zambezi. More generally, he concluded, "Lakes and dams that are man-made do more harm than good."[44] Despite this conviction, Struthers published neither photos of the emaciated buffalo nor his criticism of Kariba and other reservoirs. In the 1990s, he and other popular writers might, as well, have reported scientific findings of DDT poisoning and the mysterious "floppy trunk" syndrome affecting elephants.[45] The lake concentrated toxins in the Zambezi Valley that the river would otherwise have swept away. Of course, such news threatened to dampen readers' enthusiasm for Kariba—and, thereby, upset the all-important tourism industry. Perhaps, Struthers and others who could have alerted the public to damage caused by the lake chose discretion over notoriety.

Or, perhaps they simply didn't think about it very much. Sheer pleasure often distracted popular conservationists from the contradictions they understood too well. Gavin St. Leger begins his photographic compact disc "River Road" (2004) with images of the dam wall. "[H]ow long will you have to get out of the valley should the wall break?" he asks in the introductory text. Yet, neither the photos nor the captions that follow—documenting a journey downstream and through Mana Pools—develop this theme of engineering hazards. Only once does the text mention the rise and fall of the river "according the countries [*sic*] demands for electricity" (St. Leger 2004:3, 39). Even more than the annual rule curve, these daily, sometimes hourly, fluctuations re-engineer every moment on and along the seemingly wild river. Over beer in Harare, I asked St. Leger how he and other outdoorsmen coped with such cognitive dissonance. "You just say, 'the water is down,' " he responded. "You know why it's down, but you don't think of it. You just go on fishing."[46] Such single-minded sportsmanship relegated the dam to a pro forma mention.[47] Tourism, on the other hand, seems to provoke unbridled anger. St. Leger, like Meadows, barely tolerates groups of people and vehicles in the Valley. He takes cover in Nyamafusi, a walking safari area "away from the maddening crowds [where] you can get a taste of the real bush."[48] In person and more floridly, he excoriated the Chirundu bridge and border crossing as a "shit hole" and a "filthy, filthy hovel."[49] Truckers and traders litter the valley with detritus and with themselves, causing St. Leger—like so many others Zambezi enthusiasts—to lose sight of the concrete monolith blocking the river.

Only Cathy Buckle, among contemporary Kariba writers, breaks the literary mold. Unconventional in every way, she neither romanticizes the reservoir nor marginalizes black society from her account of it. Her novel *Litany Bird* (1999) is only set in Kariba. The story itself concerns AIDS. Jan, a white secondary school graduate, undertakes an internship at Kariba with the Department of National Parks. She falls in love with Rick, the white park warden, who soon reveals his infection with HIV. Undaunted, Jan makes love to Rick (with appropriate precautions) and cares for him as he sickens and dies.

Thus, *Litany Bird* made the case for accepting HIV-positive people—so convincingly that Zimbabwe's Ministry of Education reprinted passages in a school textbook. "Why set the novel in Kariba?" I asked Cathy Buckle over lunch in the provincial town of Marondera. Buckle knew the lake, having worked at the Fothergill Island resort in the 1980s, before buying a farm outside Marondera. "Everything was so wild and so raw . . . " she recalled, "I found it absolutely terrifying." Preferring, as she put it, not to "tell the story of glamorous animals and beautiful scenery," Buckle used the context of Kariba to make a cultural point.[50] Midway through the story, Jan and her black co-worker, Geoff, meet an elephant dying at the lakeshore. "Oh you poor thing," whispers Jan. Her pity provokes Geoff into a diatribe directed at "you whites":

> You probably grew up looking at animals as beautiful things of nature to look at and admire . . . to us an animal is . . . also the creature that raids our crops and gardens in the night and destroys in an hour what it's taken us five months through the sweat of our backs to create. (Buckle 1999:70–71)

The passage suggests a revolution in the structure of feeling encapsulated in Kariba's archive and in other modes of Euro-African expression. Could white readers learn to sympathize more with the hungry peasant than with the browsing elephant? Unfortunately, Kariba literature before and after Buckle neither raised nor answered this question. And Buckle's attention soon shifted to the violence directed against her and her farm (see Chapter 4).

* * *

Kariba offered whites an escape in multiple senses. Like any landscape, Kariba provided whites with a willing partner for their identity. In nature, they could escape from their awkwardness and downright fear amid black society. To go to nature, in fact, seemed altogether natural—so much so that few appreciated it as a flight from people. The aridity of Zimbabwe's nature, however, occasioned some awkwardness and discomfort. Thoroughly at odds with Wordsworth's values, grass plains repelled the white onlooker. At Kariba, Euro-Africans escaped from this secondary, aesthetic malaise as well. This subsequent escape differed from the first one: whites recognized it as flight. Africa's "long, grey sea" had arrived suddenly, filling the Zambezi Valley with an otherworldly presence. In traveling to Kariba, surely one left nature and Africa's typical interior altogether. This early reaction to the lake threatened to undermine the whole project of belonging. The foreignness of Kariba cast into doubt the nativeness of those who appreciated it. Authentic Africans would play in authentic African spaces. Such principles almost compelled Kariba writers of the 1970s shoehorn Kariba into the category of African nature. After flooding the valley with water, Euro-Africans flooded it with meaning and fantasy. In so doing, they ushered white readers through a third escape: from the artificial, industrial quality of Lake Kariba. With the aid of poetic text and artistic photography, observers of Kariba turned their backs on the dam wall and looked upstream. They appreciated what they saw and, without knowing it, they appreciated not seeing what they didn't see. The geometry of beauty hid as much as it revealed.

This process of authorial manipulation demonstrates the plasticity of nature—as an object of engineering and of discourse (Neumann 1998; Raffles 2002:62; Ranger 1999). Indeed, it demonstrates the similarities between building structures and building ideas. The dam builders blocked and harnessed the largest river ever before dammed. In the process, they carried out an atrocity: the dam obliterated every ecological process extant on 5,500 square km. No single project before or since has ever snuffed out this much life this fast. Yet, the lethal wall of concrete no longer causes onlookers—even romantic ones—to shudder. If engineers tamed the river, writers tamed the dam. Or, at least, conservationist writers helped their Zimbabwean readers overcome regret and accept, without guilt, a lake and all the enjoyment that it provided. Authors redeemed the reservoir. Yet their job was not easy. Just as engineers and construction workers must oppose forces of gravity and hydraulics, Kariba's writers had to remold set notions. Their texts "worked" to shift and leverage readers' preconceived ideas of nature, geology, and landscape. Some authors employed polyvalent, ambiguous symbols such as Nyaminyami. Others insinuated into their texts folk models of ecology, such as that of the divine watchmaker. The willing reader came to believe that Nature adopted the reservoir and made it her own. He or she also came to value the lake and to insist upon an ethic of care for it. Writing, in other words, transformed an instrument of technological death into a site and symbol of life. At another level, this literature bridged the gap between two conventions: the landscape of production and the landscape of leisure.[51] The same device that powers Zambia's copper mines paradoxically provides "refreshment" from industrial technology. Herein lies the true artifice of Kariba: literary and material design allowed Euro-Africans to destroy the wild and remake it in their own image—*and* to call it "wild Africa."

In this sense, white Africans took neo-European sentiments to their extreme. North Americans also naturalize and romanticize aquatic infrastructure. But their reservoirs are smaller than Kariba, and their love is more conditional. Utah's Lake Powell—completed in 1963—covers an area only one-seventh as large as Kariba.[52] Boaters make full use of the reservoir, but they have never persuaded a broader conservationist public. The environmentalist writer Edward Abbey, for instance, all but advocates a violent sabotage against the dam: through "the loveliest explosion ever seen by man" (Abbey 1968:165). More temperate critics in the United States have spawned a national movement to dismantle dams (Graf 2001; Hart and Poff 2002). Kariba, on the other hand, won over its detractors. Margaret Peach, after all, developed her "welcome home" response to the lake *after* living through Operation Noah. Only in fiction and only at the hands of the depraved do writers suggest the possibility of its destruction.[53] Why should a structure so blatantly artificial and imposed provoke such fierce loyalty from conservationists? Why, in other words, do Euro-Africans cling so stubbornly to the most patently mythological of the myths of wild Africa? Perhaps they need this myth more than do Euro-Americans. It has sheltered their identity against the insults of minority status, nationalism, and decolonization. As much a refuge as a source of refreshment, Kariba helped forestall the social reckoning whites would eventually have to face.

CHAPTER 3

OWNING LAKE KARIBA

In choosing to belong ecologically in Africa, mainstream whites undercut the established European justification for appropriating it. That conventional wisdom centered on hard work and improvement. According to John Locke, the frontier farmer "mixed his labour with" the land, enhanced it, and thereby earned rights of tenure. Previous, native occupants had, by contrast, squandered the value of the land and, hence, could no longer lay claim to it. Writing in 1690, Locke referred (erroneously) to the "uncultivated waste of America, left to nature, without any improvement, tillage, or husbandry . . . " (Locke 1980:19, 24). Meanwhile, English colonists cited exactly his notion of a "civil right" as they appropriated Indian lands in Massachusetts and later west of the Appalachians (Cronon 1983:56; Wallace 1999).[1] So-called empty land belonged to no one. Such philosophical and legal codes laid the political basis for expropriation and settlement in much of the extra-European world. Rhodesia's early land titles insisted upon improvement or "beneficial occupation," without which settlers could forfeit their land (Palmer 1977:60). The same individuals loved the pristine savannah (or strived to do so) and treated this relationship as the benchmark of integration. Why, then, should they want to transform the landscape, beneficially or otherwise? Pygmalian urges could only signal dissatisfaction and disloyalty. In other words, the logics of territorial ownership and of topographical assimilation—or, in Kosek's (2006:106) terms, "possession" and "attachment"—worked against one another. Whites could not, at the same time, argue that they belonged *to Africa* and that Africa belonged *to them*—unless they could find a substitute for Locke's notions of improvement and property.

In the event, Kariba writers outflanked Locke. They excavated notions derived from exploration, that is, from the phase immediately before settlement, agriculture, and white-led improvement. In this connection, Coetzee describes two "dream topographies": the howling wilderness and productive land use. South African writers, he argues, valued both pristine nature and human industry—without dwelling upon the contradiction between them (Coetzee 1988:6–7). North of the Limpopo, the Zambezi Valley presented writers with even better material with which to reconcile the two visions of landscape. Europeans had considered it wilderness. Then, in the late 1950s, engineers thrust their technology into the recesses of Kariba gorge. Over the next two decades, as explained in the previous chapter, writers gradually returned the valley to nature's realm. This restoration enabled a second-order form of symbolism: the re-exploration of the Zambezi—or, for those with a shorter memory, exploration for the first time. Beginning in the 1970s, writers described the valley as empty land, implying a right of discovery. Indeed, in this period, some authors acted upon the related right of naming. To name places, as Paul Carter argues in a discussion of Captain Cook and Australian history, is "to invent them, to bring them into cultural circulation" (Carter 1987:27–28). Enlarged substantially by the tourism industry, the littoral's new vocabulary invented Kariba as a white space. The Tonga, already physically removed from the southern shore, lost their place in its history. Or, rather, Euro-African authors unthinkingly omitted them from narratives of "primeval Africa." In a franker fashion, the final protocolonial discourse took black-white competition on the lake as its primary subject. In the 1980s and 1990s, Kariba writers cast blacks as abusers of the fish and the ecology of the Zambezi. Such condemnations retooled nineteenth-century notions of civilized and uncivilized hunting. Then as now, the moral failure of natives empowers whites to assume the mantle of stewardship (MacKenzie 1988). In sum, rhetorics of exploration, practices of naming, and the language of "the Hunt" metaphorically marginalized blacks from the lakeshore.

This exclusive turn supercharged the project of white belonging in dangerous ways. The mainstream of Rhodesian and Zimbabwean white writing still concentrated on inclusion. Literary and other forms of expression continued to grope toward integration. And, in pursuit of the ecological form of it, characters drove their roots into the bush, disengaging from black society and social questions altogether. Such striving presented little risk to blacks. Indeed, before

1980, whites' administrative machinery had a far greater—and more deleterious—impact on the black majority. After independence, however, the relative threat of these projects shifted. Whites, of course, lost their administrative power almost completely. Perhaps in compensation, white writers adopted a more openly political stance. The narrative of the Hunt, in particular, brought them into material opposition with African peasantry. When rural Africans fished or hunted to excess, authors not only wrote about it but also alerted and exhorted the relevant, armed authorities.[2] They sought to constrain rural black uses of the Zambezi Valley while supporting or enabling the expansion of white-dominated tourism into the valley. In print and in national policy, these authors' sensibilities carried the day (cf. Hughes 2005; Schroeder 1999). At the level of local government, on the other hand, the Tonga-dominated Binga district council advocated strenuously for the development of fisheries and fish processing. Tonga themselves fished in legal, semilegal, and patently illegal ways, sometimes citing their "entitlement to the lake as river people who had borne the cost of its creation" (McGregor 2009:156, 176). Then, Mugabe changed the rules of the game altogether. Beginning in 2000, the state abrogated conservation, suspending most enforcement and even appearing to encourage the poaching of animals. For the second time whites lost the backing of the state. Ultimately, the exclusive turn led whites into a political cul-de-sac.

EMPTY LAND AND ENLARGED SCALES

As with so much else, Kariba writers had to stretch the motif of exploration to its limits. The notion of empty land had always depended upon the willing suspension of disbelief. Cook and other celebrated explorers found "natives" living on the land. Yet those people did not "fill" the land—so the argument went—because they lived in too scattered or sparse a fashion. In this way, the eager imperialist could evacuate the land rhetorically. Could a white writer make the same argument with respect to Lake Kariba and the littoral? Rising water had certainly forced the evacuation of mid-Zambezi Valley Tonga. Yet, those 57,000 people figured in all the historical accounts of the dam. Even if they lacked sympathy, whites did not altogether forget Kariba's human cost. Indeed, anglers valued the submerged remains of huts and granaries. These structures provided ideal habitat for certain sport fish, and most anglers used depth finders and/or antique, large-scale maps to locate them. They could hardly represent the land as empty or even sparsely populated as Cook and other classic explorers had. Rather, the interracial shadowboxing promoted weaker and less direct claims. Writers suggested that the Tonga did not *know* the impounded Zambezi Valley. In a second, even more oblique move toward appropriation, white writers jumped scale. They enlarged the frame of reference within which the mid-Zambezi Valley lay. If 57,000 Tonga seemed significant along 300 km of river, they seemed less so over the entire 2,700 km course of the Zambezi. In effect, expressions and images raised the denominator in the ratio of person to land area. They suggested a landscape that was, *on balance*, uninhabited. Writers would have immediately denied such an absurdity, but their metaphors suggested otherwise. Just below the threshold of writers' and readers' consciousness, Kariba literature after 1970 *emptied* the mid-Zambezi Valley.

Exploration and implied emptiness appeared most subtly in narratives of descending the river. In print, David Livingstone had last run the Zambezi during a long expedition between 1858 and 1864. In the intervening century, colonial boundaries and the related priorities of political control and settlement had curtailed such treks. When whites returned to equivalent adventures, they drew ready inspiration from the past. *Zambezi Odyssey*, published in 1974 by Stephen Edwards, begins by quoting the early twentieth-century hunter Denis Lyell: "It will be years before the love of a wild, wandering life and the spirit of adventure disappear from the Anglo-Saxon race, but in years to come there will not be any wild countries left" (Edwards 1974:1). Fortunately, as Edwards continues, "in Africa, there do remain vast tracts of land barely touched by civilization and one can embark, for a while, on a simulation of the old life—unregulated, independent, fending for himself" (Ibid.:2). On a farm north of Harare, Edwards boarded his canoe and paddled down the Umfuli River, into the Sanyati, across Lake Kariba, and down the Zambezi to the sea. He endured tremendous hardship, aching hunger, and moments of sheer terror. Luckily, Africans emerged unexpectedly from the bush to help him portage around the most lethal rapids. As in the "old life," African labor and periodic rescues do not seem to dispel the sense of maverick frontiersmanship. Lake Kariba itself bores Edwards. He finds it barely tolerable and quickly paddles to the Mana Pools, "where nature is still unspoilt" (Ibid.:136). In this sense, Edwards takes exception to the naturalization of Kariba already underway in 1974, but his adventure narrative propels white thinking into the next, appropriative phase.

In conversation, Edwards articulated his concept of exploration in even stronger terms. Meeting at an airport bar outside Kruger National Park, we turned immediately to Edwards's adventures. Before and after the Zambezi descent, he had put his paddle in numerous southern African watercourses. Now, in the comparative comfort of a remote South African farm, he was working on a second book that would rebut the statement by Hastings Banda, Malawi's deceased life president, that whites did not discover anything. For the sake of argument, I tried to articulate Banda's position: white "explorers" encountered native Africans wherever they went, and those people surely knew of the closest river. Yes, admitted Edwards, "they may have known it was there, but they had no frame of reference, no latitude and longitude . . . They covered very small home ranges, and their curiosity did not extend to asking, 'where is the source; where is the mouth.' "[3] Edwards's answer to me and to Banda neatly retraced the most elegant of empty land rhetorics (Carter 1987:64). Although aboriginals used the Zambezi and others rivers, they did not know them as *rivers*—that is, as water's long march from airborne impact to ocean dumping.[4] Even the most local riparian segment meant nothing to its inhabitants in global terms. Tonga, in other words, could not have found their villages on a Mercator projection. Such dismissals of local knowledge could indicate a deeper disrespect. In 1993, Edwards's employer, the hunting firm Mozambique Safaris, had chained suspected poachers to a tree for three days, a notorious act of vigilantism that sparked the movement toward community-based wildlife management in Mozambique (Hughes 2006:173). Edwards did not wish to discuss the incident, and he may have played no part in it.[5] Yet, in other ways, he expressed a similarly misanthropic or Malthusian approach to conservation. "Unless something comes along that is infinitely more effective than AIDS," he predicted, "the whole planet is buggered . . . We need a removal of 2–3 billion people."[6]

Although no similar descent narratives appeared for ten years or more, the genre flourished in the late 1980s and early 1990s. In 1988, David Lemon published his account of a journey across Lake Kariba apparently as harrowing as Edwards's odyssey. Born in Britain in 1944, Lemon emigrated to Rhodesia in 1963, where he served in the police force of Kariba town. In 1985, he rowed 560 km across the lake in both directions at the long dimension. He wrote a journal-like account of his voyage—*Hobo Rows Kariba*[7]—and framed it as the discovery of "a largely unexplored shoreline."[8] I met the author in Britain and asked him how he could "explore" an area from which Tonga were expelled. "Tonga activities don't count," he replied, "because the shoreline is new."[9] The dam, he implied, extinguished Tonga claims to the pastures, fields, and hunting grounds it destroyed. Without evident contradiction, then, *Hobo* expresses the author's "antipathy towards domestic animals in wild places" and his "fears about cattle encroachment," as the eradication of the tsetse fly opens up the Zambezi Valley to livestock.[10] I asked Lemon to account for this vehemence. "It was just the thought of more civilization encroaching . . ." he responded[11] Such anxieties, of course, left little room for the Tonga and others to make a living from the littoral. They even crowded out many whites. "Why can't visitors just enjoy the peace of Kariba?" Lemon asks in *Hobo*, having just "passed a large houseboat [whose] . . . generator did nothing for my mood" (Lemon 1997:137). The entire lake generates electricity, I noted in our interview. "It just doesn't make noise," retorted the author. Slightly rattled, Lemon continued in a proprietary vein: the tsetse fly "keeps my little bit of Africa for me."[12] Once expressed so plainly, that sense of entitlement embarrassed the Lemon. *Hobo* elides any such ambivalence. "As a rule," the author wrote, "I feel instant antipathy for tourists on *my* lake" (Lemon 1997:111; emphasis added). Tourists, Tonga—Kariba's rower would dispossess them all.

Set against this bluntness, subsequent descent narratives excluded Africans in far more subtle ways. Perhaps the greater comfort and safety of these explorations softened claims of entitlement. Over a number of years in the late 1980s, Jumbo Williams and Mike Coppinger—boarding school buddies from Harare—piloted a motorized craft along various sections of the river. In 1991, the two published the photo book *Zambezi: River of Africa*, and Williams shortly emigrated to the San Francisco area. We met at his house for dinner, where conversation plunged into the book's direct and indirect language of discovery. As is conventional, *River of Zambezi* begins with a justification for the trip: Livingstone had preceded them, the authors write, but "[t]he full 2700-kilometre length of the waterway had never been comprehensively explored" (Coppinger and Williams 1991:13). (War in Angola and Mozambique, in fact, foiled the authors' attempt—despite assistance from the South African army in Namibia and Stephen Edwards himself on Lake Cahora Bassa.) I questioned Williams on a later, even more striking mention of Livingstone. The authors credit him with hunting the first puku, a rare antelope found in the Luangwa Valley (Ibid.:139). Williams clarified:

yes, Africans had shot and snared puku long before that. Livingstone, however, obtained the "first specimen recorded" of the species, thereafter known as *Kobus vardonii* Livingstone 1857.[13] Designed by Linnaeus a century before that, taxonomic nomenclature recognizes not the first person to hunt, kill, or even name an animal, but the first person to enter it into a global grid. Here, taxonomy played the same role as did cartography in Edwards's view: imperial systematics effaced local knowledge and local people. Elsewhere in *River of Africa*, Coppinger and Williams exclude Africans even more explicitly from the landscape. The river just above Lake Kariba, they write, "really belongs to recreational fishermen" (Ibid.:94). Again, Williams in our conversation clarified: he had encountered some Tonga fishing for subsistence, but (white) angling "is the majority activity."[14] Surely, a more scientific census would have unearthed the Tonga communities living along the banks.[15] Even if plainly erroneous, though, the claim represented a suppressed desire: to find oneself not outnumbered, but at last outnumbering Tonga and other blacks. Did a home for whites nestle in the Zambezi Valley?

The final descent narrative approaches this possibility in an even more roundabout fashion, using a rhetoric of time. In *Zambezi: Journey of a River* (1990), Michael Main shifts the river from history to geology.[16] Born in Zimbabwe, he did not actually run the river continuously. Still, the narrative *Journey of a River* begins at the source and ends at the mouth. Like the other descent authors, Main wished to put a broad range of political and economic phenomena out of his mind. Or, at least, he sets these aspects of the present in a framework that makes them appear insignificant. Jumping scale in time as well as in space, the book opens with a predictably geological history of the river from 200 million years ago to the present. To explain Victoria Falls, writes Main, "it is necessary to go back to Triassic times . . . when the supercontinent of Gondwanaland was still united . . . " (Main 1990:9). In a similar geological vein, *Journey of a River* discusses and definitively refutes the paleo-lake myth regarding Lake Kariba (see Chapter 2). Recall that Main's earlier work on the Kalahari—discussed in Chapter 1—traces the Magkagikgadi paleolake over the past several tens of thousands of years. In his current home in Gaborone, I asked Main to account for his passion for deep time. Geology, he responded, "gives you context in a world sense, in a truly global sense." Geology also marks chronological precedence. "Everything else sits on top of it," Main continued "People, politics, culture on top of geology." Such notions of *longue durée* reduced Africans and African nationalism to the status of mere transient events. Main's rhetoric suggested not that Africans should be wiped out, but that they had never really mattered. Ever so delicately, this dismissal promoted ecological belonging and social escape—not least for Main himself. "I'm not interested in politics," the author confessed. In geology, "I have a kind of place to fit in."[17] A light-skinned African could belong in the Triassic, and—although Main did not say it—blacks did not. The temporal scale suggested not merely a white home in the Zambezi Valley, but a white epoch.

Names and Brands

To appropriate the landscape symbolically, one had to do more than evacuate blacks from it. Government trucks had moved the Tonga physically. Spatial and temporal scales associated with exploration pushed them metaphorically to the margins. But these tropes did not automatically replace blacks with whites as river people. How, then, could whites positively assert the primacy of their identity and heritage with respect to Lake Kariba? Again, the most obvious means—a celebration of the dam—lay off-limits. In a more convoluted fashion, whites would have to link themselves with the (imagined) wilderness of what was once the mid-Zambezi Valley. To do so, they turned to a practice associated with, but distinct from, exploration: naming. Names filled a landscape with meaning—or, even more profoundly, made it possible for the land to hold meaning at all. "It was the names themselves," writes Paul Carter of "Botany Bay" and other Anglo-Australian neologisms, "that brought history into being, that invented the spatial and conceptual coordinates within which history could occur" (Carter 1987:46). A colonial name replaced and effaced its native analog. Or—better still from the settler's perspective—the new name recategorized the topography, breaking it into units and features that made earlier meanings unrecognizable. Kariba presented writers and other whites with an ideal opportunity to employ such linguistic technology. In the Zambezi Valley, rising water created topography entirely new to inland southern Africa. In fact, topographical change necessitated naming at two levels: designating specific places and designating the kinds of landforms that had suddenly appeared. For the most part, writers did not personally invent the names for new islands, bays, and so on, but they propagated all of them in tourist and photographic literature. Such travel-oriented writing also assisted in a more commercial technology of naming: the branding of certain sites and aspects of the lake for marketing and consumption. As the term "branding" implies, all of these names indicated ownership: haphazardly and without explicit intention, whites marked Kariba as their own.

With varying degrees of directness, place names coupled the Lake to white history and culture. As the clearest example, Fothergill Island recognizes Rupert Fothergill, hero of Operation Noah. Other inlets carry Tonga or African monikers, but simply because the Rhodesian Mapping Committee applied the name of the tributary that empties into them. Operating until 1980, this official body did more than any other individual or group to mark the lake as white space. As hotels went up south of the dam wall, the Mapping Committee labeled the adjoining shoreline Leisure Bay. In a similar association between white lifestyles and land features, the committee recognized air force squadron names. During the guerrilla war of the 1970s, Quickstep squadron covered a tiny speck of an island, now known as Quickstep Island.[18] David Lemon describes a similar, if unofficial, wartime act of naming. *Never Quite a Soldier*, the memoir of his police service, describes patrols on Lake Kariba and his filthy camp located midway along the lake. As Lemon recalls, a visiting boatman disembarks with the sarcastic insult, " 'back to bloody Paradise Island.' " "The original Paradise Island," Lemon explains, "was a holiday resort—long loved by Rhodesians—off the Mozambique coast. Now we had our own Paradise and the name stuck" (Lemon 2000:88). This name, then, performed two rhetorical moves. It anglicized a site in the lake, suggesting that whites belonged on it. Secondarily, the appellation compensated Rhodesians for their loss of access to a place where they apparently no longer belonged: shortly after its independence in 1975, Mozambique closed its border to all Rhodesian holidaymakers. Rhodesians, in other words, claimed Kariba as *their* paradise, and the name *did* stick. Through such christenings, whites filled the lakeshore with their recreation, their history, and their fantasy.

Geographical terms served the same purpose. These *metanames* for categories of landscape features solved a problem described by Coetzee (1988): the mismatch between European language and extra-European land. From California to Australia, dry climes have forced colonizing Anglophones to borrow and improvise vocabulary.[19] In southern Africa, to this day, English-descended Africans employ Bantu-language or Afrikaans words for features of the bush unknown in Europe. African landscapes, in other words, continually Africanize English—and undercut Euro-African claims to belong on them.

As an artificial landscape, Kariba offered a means of reversing this process of alienation. Again, however, the linguistic maneuver, required grace. New labels would need to carry meaning without reference to the dam that made them necessary. In the event, writers borrowed from their glacial heritage. In a coffee-table book of the early 1980s, Pitman describes the Sanyati Gorge as a "fiord"[20] (Figure 3.1). "Loch," would have done just as well, he told me.[21] Although neither is English, both terms recall the last ice age and the rugged, northern European coastline it created. Could Kariba's writers not find descriptors indigenous to Zimbabwe? I posed the question indirectly to Kenmuir, who in a novel also refers to the "fjord-like" Sanyati (Kenmuir 1993:57). "I've never actually seen a proper fjord," he explained, but, based on pictures, "there's not an African word that would describe that."[22] Such lexical practicality barely concealed a powerful, if contradictory, claim: speakers of Tonga, Shona, or Ndebele could not describe (or understand) the new landscape. Euro-Africans, equipped with linguistic vestiges of a terrain left behind, could now do so far better. Who, therefore, belonged more on Kariba's shore? Language, in other words, displaced blacks and allowed whites to occupy the indigenous position.

Figure 3.1 Sanyati Gorge, Lake Kariba, photograph by author, 2003

What names and metanames accomplished elegantly and subtly, brands effected more brashly. Through an expanding set of commercial logos, the white-run tourism industry virtually patented Kariba. Jeff Stutchbury—described by Coppinger and Williams (1991:133) as a "grand old Zambezi character"—set this appropriative dynamic in motion. Born in Britain and Zambia, respectively, Jeff and Veronica Stutchbury came to Matusadona National Park in 1970. There they managed tented camps, as the photo book *Spirit of the Zambezi* (1992) documents. While based in Matusadona, Jeff studied the shoreline and—in a bit of empty land discourse—took "photographic forays into the unknown and unexplored bays of the Ume River" (Stutchbury and Stutchbury 1992:17). Then he hit upon a "masterful idea . . . the water wilderness safari concept . . . [including] a totally water-orientated safari base camp" (Ibid.:17). Water wilderness relied upon a circuit of shoreline platforms floating and suspended in trees, between which the tourist traveled by flat boat. Disseminated in brochures and advertisements, the moniker also relied upon one's willing suspension of disbelief. After all, the water was the least wild element of Kariba. In Harare, I interviewed Veronica Stutchbury, who had written the text of their joint book. She admitted the irony in a notion of wild water but emphasized the couple's need for a label: "How do we sell this thing? How do we explain this?" she remembered them asking each other.[23] Their formula worked, and, as Veronica wrote, Jeff "made his mark all over the shoreline . . . " (Ibid.; cf. C. Williams 1979). Indeed, the couple later established the widely respected Chikwenya Camp on an island in Mana Pools National Park. After Jeff died in 1992, however, Veronica left the safari business altogether. "Why?" I asked. One becomes, she answered, "incredibly possessive of the countryside and of the area."[24] Hers was an unusual misgiving.

Amid the tourism boom of the 1990s, advertisers and other image-makers—white and black—abandoned their qualms and even their good taste. They turned Nyaminyami into a label. From the outset, whites had popularized the Tonga river god. In the late 1950s, as engineers battled against unprecedented floods, Nyaminyami signified wild Africa—fighting for her freedom—only to be subdued by white civilization (see Chapter 2). In the ensuing decades, that respectful, if obviously romanticized, discourse yielded to a more offhand treatment, epitomized in Rex Taylor's contributions to the angling magazine *Zimbabwe Fisherman*. His "Kariba" column focuses on storms and boating through them. Nyaminyami stirs water and wind and, as he wrote in 1993, "is waiting to gobble you up!" Two years later, a westerly hit the annual regatta, but skippers made land safely. "Nyami Nyami, stand aside," reported Taylor triumphantly, "you were beaten once again!" (R. Taylor 1995:33). I met Taylor at his house overlooking the lake and asked him if he believed in Nyaminyami. He responded with a rhetorical question and an admission: "Who do you think blows up the storms when you want to go sailing? . . . You've got to blame it on something. It's a good feature."[25] Nyaminyami, in other words, served as a literary ornament. The river god also served as a literal ornament: curio makers carved coiled snakes as fast as gullible visitors could buy them up. In the 1980s, three Tonga chiefs sued the sculptor Rainos Tawonameso for copyright infringement.[26] They lost the case, and, by 2000, white and black entrepreneurs imprinted river gods on shirts, shorts, and jewelery, doing particularly brisk business around Victoria Falls (McGregor 2009:165). Royalties flowed to the Nyaminyami Corporation, a Kariba-based partnership that held the logo's patent.[27] In 2002, the newly formed Kariba Publicity Association adopted the river god as its emblem. By then, the objections of devout Christians caused more concern than those of Tonga chiefs. Over lunch at the Cutty Sark Hotel, I asked a board member of the association whether it had stolen Nyaminyami. "Nyaminyami is not for one person," he responded gamely, "Nyaminyami is for everyone."[28] Tonga would have to share their god.

This appropriation of names and symbols reached a crescendo in the hotel industry. Although black himself, the board member Kennedy Matarisagungwa had actually represented the two hotels whose names most suggested a white space at Kariba.[29] In 1986, after 15 years in the hospitality industry, he took over food and beverages at the Caribbea Bay Resort and Casino. Still at lunch overlooking the lake, I asked him what the name could mean. Matarisagungwa recalled the founding of the hotel in 1974. "That's a Caribbean sort of thing because of the buildings." The investors, it seemed, exploited Kariba's phonemic resemblance to a New World sea. Yet, guidebooks and the hotel's own literature advertised "Sardinian architecture" and "Mediterranean-style *casitas*" (Funnekotter n.d.:3; Martin 1995:83). Did the hotel denote the eastern or the western extreme of the Atlantic? Matarisagungwa didn't know, but he assured me that the founders "wanted something different from Zimbabwe, something original."[30] If Caribbea Bay's precise cultural referents somewhat escaped him, Matarisagungwa grasped the Cutty Sark rather better.

He had taken over that hotel as general manager in 2004. Again, I enquired about the name—related, of course, to the Cutty Sark clipper that flew the Union Jack in the mid-nineteenth century.[31] In the 1960s, he remembered, an English family had bought the property and planned to dock a replica of the ship in front of it. They never realized that dream, but, as Matarisagungwa showed me, the bar contained a smaller model as well as maps showing the Cutty Sark's global peregrinations. Only the previous week, the real ship moored in Chelsea had suffered extensive fire damage. With genuine concern for the vessel and her legacy, Matarisagungwa summarized his bar's didactic purpose: "The message there is the history of that ship—that it used to go from that point to this point and that it was the fastest ship."[32] Our thoughts had traveled a long way from Kariba.

Such transporting exoticism gives Kariba an air of the unreal and of the hyperreal. Fjords, river gods, and *casitas* conjure multiplex illusions. Even the writers and promoters who invoke these terms recognize how little they understand them. At the same time, Kariba embraces a hyperreal aesthetic that, in Umberto Eco's sense, strives for greater authenticity than the original. In Disneyland and Hearst Castle, for instance, Eco criticizes California's "search for glory via an unrequited love for the European past." By the end of the twentieth century, Kariba's tourism industry was remaking and marketing the same heritage—at a cost. As Eco continues, "to recreate Europe in desolate savannahs destroys the real savannah and turns it into an unreal lagoon" (Eco 1986:28). Matarisagungwa's white boss drove this point home. Brian Keel, managing director of the hotel's chain, greeted me upon my return to Harare.[33] His family had bought the business only a year before, having lost their farm outside Marondera (close to Cathy Buckle's, in fact). Still, Keel played Kariba's game of symbolic expropriation like an old hand. First, he complained about locals along the lakeshore: "That fishing in front of the hotel spoils my wildlife." Such people were, in fact, trespassing on the strip of hotel-owned lake frontage in Kariba town. To restore littoral biodiversity, he proposed a more radical form of zoning: demolishing the Nyamhunga township and rebuilding it inland and uphill.[34] Along the shoreline and closer to the hotel, he would construct another kind of settlement: "a true Batonka village." Guests would sleep in huts, Keel elaborated. "It would be a Batonka hotel, for argument's sake."[35] Or, for argument's sake, it would be a hyperreal copy of people whose flesh-and-blood-and-fishing he and other claimants to

Kariba resented. But, the welter of desires and daydreams was too tangled for facile criticism. Nostalgic for older, northern coasts, whites *did* own the "unreal lagoon" they made in their minds.

HUNTING, STEWARDSHIP, AND OTHER MORALITY PLAYS

Fantasylands require heroes and villains. In the final appropriating discourse, Kariba writers cast these roles along racial lines: with some exceptions, the good guys were white and the bad guys were black. This Manichean turn took Kariba literature into unfamiliar, sociological territory. Yet, it also replayed a long-standing narrative of the Hunt. From the Middle Ages, English depictions of the countryside extolled elite sportsmen, as against grubby peasants looking for a meal (MacKenzie 1988:15). The former rode to hounds, giving the fox a chance of escape and, at the inevitable denouement, killing it swiftly. The village poor, on the other hand, set snares, in which animals died slowly and painfully. In short, landed gentry hunted forthrightly, precisely targeting individuals of a given species while tenants used a cowardly, impersonal device known to destroy animals indiscriminately. By the eighteenth century, elites invoked this hierarchy of values to justify enclosing the rural commons and evicting tenants. Later still, the sporting class employed the same trope in reserving tracts of British African savannah for white hunters and against African herders and farmers. Rhodesia improvised little in this regard, and independent Zimbabwe perpetuated both its protected areas and the stigma against subsistence hunters. Yet, by the 1990s, Euro-Africans sensed a slackening of zeal in enforcement—most keenly felt when the Department of National Parks and Wildlife Management purged its white staff in 1994. Perhaps reacting to this sudden spectator status, Kariba writers adapted the narrative of the hunt to Kariba's lacustrine environment. In pursuit of fish, idealized whites demonstrated self-restraint, mercy, and ecological wisdom while blacks succumbed to atavist urges, cruelty, and improvident natural resource management. Of course, no writer emphasized race crudely or gratuitously. Yet, all presented light-skinned characters as indispensable conservationists. Dark-skinned ones, at best, played a supporting role.

Dale Kenmuir's fiction sketches this morality play in its clearest form. *The Catch* (1993), which nearly won an award for youth fiction, takes place during the annual Kariba International Tiger Fishing

Tournament.[36] Among its participants, virtually all white, two local residents—a father and son struggling to make ends meet in the kapenta business—hunt the record-breaking fish.[37] Through superior ecological knowledge, they catch their prize, but the son confesses to having improperly baited the water. Dad throws the fish back. Later, "Prof [the boy's scientist friend] stared at us in some surprise," writes Kenmuir, narrating in the voice of the boy but clearly modeling the Prof on himself. In disbelief, Prof questions, "You deliberately threw away your chances . . . of having security for the future, because you preferred not to win by cheating?" (Kenmuir 1993:95). In our conversation, Kenmuir portrayed whites as requiring and deserving such security. Under black rule, "it's always been total uncertainty," he explained, "it's like living on the edge of a volcano."[38] Perhaps their abundance of good deeds will win whites a reprieve from danger? Another of Kenmuir's novels fulfills this wish dramatically—all the more so because evil wears a black face. *Dry Bones Rattling* (1990) follows the career of One-Eye, a horribly evil fish-netter. Known locally as Nyoka—meaning "snake" in Shona—he lives up the Sanyati River and, like Conrad's Kurz on the Congo, has gone mad. By pretending to be a powerful witch, Nyoka terrifies his Shona-speaking labor force into submission. He and they string nets across the Sanyati and kill schools of spawning bream and tiger. To save the day, two white boys and their trusted African sidekick track Nyoka. At great peril, they find him, win over his labor force, and pursue him until he falls into the Sanyati to be devoured by crocodiles. In Kenmuir's sequels (1987; 1991), both boys become noted park rangers. Thus, the ethics of the colonial Hunt prevail, and so does a postcolonial Euro-African fantasy. Nature punishes the black poacher with death and rewards the white outdoorsman with professional success. Kariba does what the black-ruled state fails to do.

Nongovernmental organizations also began increasingly to fill the gaps in state action, especially with respect to racially coded environmental stewardship. In 1992, Lis Dobb, of the almost all-white Wildlife Society of Zimbabwe, wrote in *The Fisherman* regarding the dangers of siltation. Agriculture everywhere was causing erosion, but the black communal lands contributed more than their fair share by hectarage, threatening Kariba and other dams. Rather than hewing closely to these agronomic facts, however, Dobb reached out emotively to the magazine's white readership. "The Wildlife Society of Zimbabwe," she concluded, "is trying to save your leisure-pleasure" (Dobb 1992:33). Even black-run agencies—which in the 1990s took over the mainstream conservation movement—could unwittingly nurture white fantasies and interests. In 1994, for instance, the Harare office of the International Union for the Conservation of Nature (IUCN) distributed a calendar poster intended to express sympathy for displaced Tonga, but which subtly conveyed exactly the opposite sentiment.[39] Under the title "social perspectives on natural resource management," the text emphasized IUCN's shift from fortress conservation to community-based approaches. Puzzlingly, the calendar juxtaposed this small-is-beautiful sentiment with a satellite image of Lake Kariba. Surely, the vantage of space, from which no people were visible, undercut the message of "social perspective" and "community." At a workshop, I put this challenge to Joseph Matowanyika, who had worked at IUCN, Harare at the time. Having approved the calendar's design in 1993, he recalled how it encapsulated "hard science reflecting the impact of human activities."[40] Indeed, upon closer examination, the satellite image *did* show Tonga people: as red "fire scars."[41] Lake Kariba itself warranted no alert of this kind. On the contrary, the text dignified the reservoir as a "natural resource." In this contradictory fashion, IUCN's poster suggested a moral hierarchy equivalent to that of the Hunt: (black) swidden farmers disfigured the landscape while (white) aquatic recreators did no harm. Even under black, professional direction, conservation conveyed implicit prejudice.

The political jolt of 2000 brought conservationist whites to the fore again—along with their own growing anxiety. ZANU-PF squads roamed Kariba town, attacking and terrorizing supporters of the opposition. Addressing a related problem, *The Fisherman* gave space to the Zimbabwe Conservation Task Force, a fledgling, private organization. Armed groups, the Task Force reported, "are seen netting daily in prohibited areas [of the lake] and when questioned by National Park rangers, the rangers were held at gunpoint."[42] At Kariba, law, order, and decorous hunting seemed to be falling apart. "In most of the dam, there is no fish now," accused Glen Powell, one of the instigators of the Task Force.[43] Almost beside himself, Powell told me over (many) drinks in Harare of the decay of Zimbabwe's natural resources. Then, he confided his plan for an escape and salvage operation. He would replicate Kariba reservoir in friendlier territory. He would buy 10,000 square km of land in a neighboring country and translocate to that enormous zone central African gorillas. This exotic species distracted us from the tiger fish. I noted that African savannahs do not contain gorillas' preferred ecosystem of moist, montane woodland. Powell had anticipated this problem. "If you create a forest with the right habitat," he assured me "the gene pool will do what it has to do, and all we have to do is safeguard the wildlife free of human intervention."[44] I pressed on: How would he create a forest? "We pump water," Powell directed, probably from the Zambezi and possibly generating hydropower as well. The Kariba discourse had come full circle. As the Zambezi's first reservoir deteriorated, Powell was proposing to manufacture a duplicate—complete with its unlikely blend of engineering and wilderness. Would narratives of appropriation repeat themselves? Presumably, local residents would have to make way for the reservoir and the consequent private preserve. Water would wash away their history and their claims. Tourists could then "explore" the empty land, and managers would find ways of exercising effective stewardship over the kidnapped gorillas. I could see new literary challenges stretching decades into the future. I also saw an immense zoo, and I told Powell so. No, no, he objected, "Once you interfere [as in a zoo], you've got a problem. You should let life take its course."[45] Like Kariba, the imaginary hydro-engineered habitat would naturalize itself, and whites would own it.

* * *

At Kariba, postindependence white conservationists doubled a risky bet on belonging. From the beginning, the lake had slaked Rhodesia's thirst for a waterscape reminiscent of glaciated Europe. Rhodesians responded to the lake with an aesthetic joy far surpassing its electric services. Into Kariba, they thrust their cameras, boats, and fishing rods. But was the landscape African? Yes, by the 1970s, imaginative writers had shifted the lake from the field of technology to that of nature. They depicted a "water wilderness" of primeval, prehuman Africa. In a second, more political shift, this bit of fantasy added cultural value to entertainment: if whites felt at home on Kariba and Kariba contained deepest Africa, then whites truly *were* African. Boating, in other words, conferred belonging—as long as one suspended disbelief in numerous ways. Rather than recognize the fragility of this foundation, though, white writers and popular conservationists built further layers of association upon it. Belonging, they insinuated, ran in both directions, such that the landscape—to which whites belonged—also belonged to whites alone. This monopolistic corollary suggested that the Tonga and other black residents of the mid-Zambezi Valley did *not* belong there. How could even willing white African readers accept such an exclusion as at all credible?

Again, imagination lifted a heavy load. The triple language of exploration, names and brands, and morality plays pushed the Tonga masses to the valley's edge. In their place, a handful of white fishers and boaters knew the lake best and could best protect it. Such traffic between nature and race has become conventional in the neo-Europes. How odd that it should arise and persist in a failed neo-Europe: white Zimbabweans acted with the hubris of a nation-state when the caution of an enclave society might have suited their circumstances better. But that misjudgment only became apparent later.

In the meantime, tourism and conservation catered to aesthetic values rooted in the Rhodesian experience. In the 1990s, promoters began applying the term "ecotourism" to describe the confluence of environmentalism and hospitality. It became a global marketing fad. Rather than broadening the scope for enjoyment of Zimbabwe's savannah, though, that merger narrowed access to it. At Kariba and elsewhere, ecotourism has not validated and celebrated landscape as such. Indeed, tourism and conservation both strive for an ideal countryside. In both fields, proponents wish to preserve flora, fauna, and land forms as artifacts of an earlier age. The most naïve identified that past period as prehistory, the Pleistocene, or some other prehuman time of innocence. Such fancies emptied the landscape of its people. Yet, they could not explain a stubborn detail: the tourists themselves. Perhaps in response to this dilemma, shapers of North American wilderness aesthetics have constructed a more historically specific scenario. Outdoor sports, for instance, subtly "invo[ke] . . . histories of European exploration and adventure" (Braun 2003:183). More explicitly, national parks of the U.S. West stage an aesthetic of nineteenth-century expansion (Louter 2006:127). In Africa, artificially empty parks have played to the same frontier nostalgia with perhaps less awareness (Neumann 1998). Only at Kariba did the ironies threaten to overwhelm wilderness myths. A coastline built in the 1960s hardly recalled nineteenth-century pioneering—especially since that pioneering took place far inland. Again with little awareness, hoteliers and advertisers reached back to an earlier era. To visitors, they presented what Denis Cosgrove describes as a Renaissance "tropicality": the aesthetics of "a landscape viewed from the imaginative distance offered from on board ship" (Cosgrove 2005:205). This maritime vision relocated Kariba to the Caribbean, as an Edenic archipelago best enjoyed by watercraft—and best enjoyed by whites.[46] Black elites, of course, now frequent the hotels and casinos, but they seem less moved by the lake itself. As of 2008, no black

Zimbabwean has published text or photos of Kariba. It remains as ecotourism shaped it: coded white.

If conservation, then, has been good to Euro-Africans it should be extended to blacks. There are two means of neutralizing any structurally racist practice: to abolish it or to generalize it. In Zimbabwe, poachers are currently pursuing the former option. In 2003, Harare's *Financial Gazette* reported massive poaching in protected areas and privately owned conservancies.[47] Even if overstated, such concerns indicate a change in the balance of political forces regarding wildlife. After independence, the state and private enterprise continued to propagate what Thomas McShane of the World Wildlife Fund calls the "myth of wild Africa" (Adams and McShane 1992). Like McShane, many conservationists and even Pitman's Zambezi Society—which in 1998 commissioned a study of "perceptions of wilderness"—recognized this folly.[48] Yet, it seemed necessary to practice the deception. "Pristine Africa" lured Europeans, North Americans, and their hard currencies to Zimbabwe. No longer: as political order collapses, tourists fear to come. Meanwhile, the state increasingly neglects wildlife. The compromise of 1980—wherein white fantasies funded a black government—lies in shreds.[49] If they act quickly, however, conservationists can save some aspects of nature by striking a second, truly postcolonial bargain. That bargain would apply the insights of Kariba to democratize conservation. If the state has, in the past, protected white cultural heritage, it can also protect black cultural heritage. Motivated authors and photographers can as easily redeem (black) cattle pastures and maize fields as they have a (white) hydroelectric dam.[50] Europe itself provides a model. There, national parks allow farming and settlement. In Africa too, there is space—in conceptual terms—for agro-industrial activities on land naïvely called "wilderness." If they do not make that space, then admirers of Zimbabwe's wildlife may well lose it.

Part 2

The Farms

CHAPTER 4

HYDROLOGY OF HOPE

The process of appropriation moves from diffidence to entitlement—and sometimes back again. At first, settlers and colonizers ask themselves, "Do we belong here?" Over time, such doubt may dissipate—as it did in the United States. That country occupies an extreme position among territories colonized from overseas. Whites achieved demographic, political, and economic dominance, securing the United States as a "neo-Europe" (Crosby 1986:2). Zimbabwe lies at the other extreme—among what one might call failed neo-Europes. Having conquered the territory in the 1890s and alienated the fertile highveld in ensuing decades, whites never approached numerical superiority vis-à-vis native peoples. Stuck in this "demographic conjucture," whites' population never exceeded 5 percent of the national total.[1] In agriculture, at the end of the twentieth century, almost 4,500 families of white commercial farmers controlled roughly 33 percent of Zimbabwe's surface area—in a nation of 12 million (Figure 4.1). Whites then had reason to feel what Wagner (1994:171) describes, in South Africa, as an "emotional and moral unease with the fruits of conquest." Still, white farmers, who mostly survived an incomplete land reform in the early 1980s, displayed an almost Euro-American degree of confidence—one totally unwarranted by political trends.[2] In the 1990s, whites ignored warnings of a more thorough land reform. In 2000, when paramilitary bands occupied their land, farm owners reacted with shock and disbelief. Unprotected by the police and frequently barricaded in their houses, they still felt that

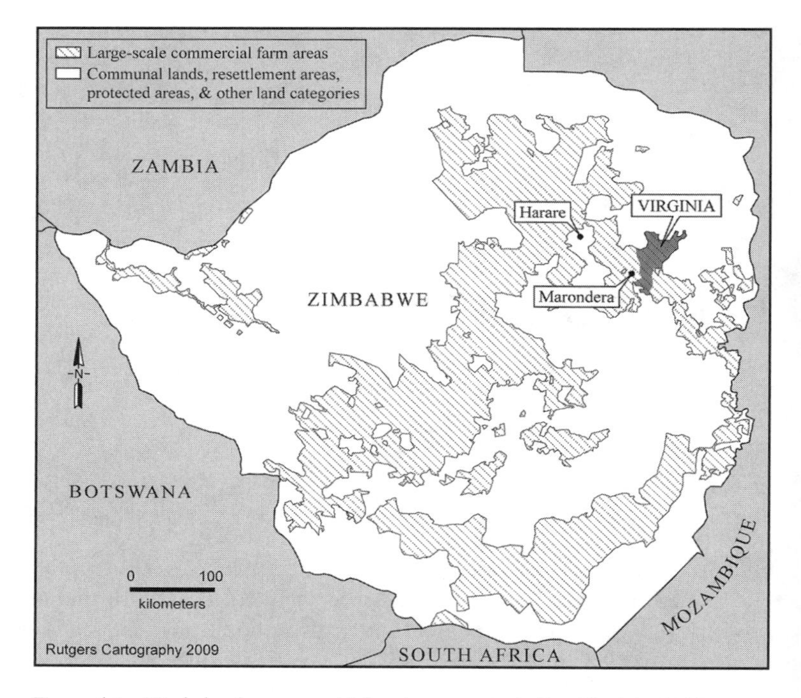

Figure 4.1 Zimbabwe's commercial farming areas, including Virginia, 2000
Source: Map by Michael Siegel, Rutgers Geography Department.

they *belonged*, as owners, on the highveld. How could they—indeed, how could any European-derived minority—develop such a resilient claim to extra-European territory?

In large part, southern African whites did so by idealizing, celebrating, and generally obsessing about the territory itself. Of course, agriculture inevitably brought farm owners into frequent contact with farm workers. "The labor" could facilitate or disrupt farm operations, enriching or infuriating the boss. Even so, these blacks operated within the confines of whites' administrative project: Euro-Zimbabweans *managed* them but did not construct an identity around them. Commercial farmers, like many other savannah whites, felt the primary tension or contradiction as (white) Man against the land—not white against black (cf. Krog 2003:76). If European-descended farmers could only master African land, they presumed, then all else would fall into place. In this effort, they acted on Coetzee's second "dream topography," complementing the first literary vision of empty land (see Chapter 3). This imaginary landscape constituted:

> a network of boundaries crisscrossing the surface of the land, marking off thousands of farms, each a separate kingdom ruled over by a benign patriarch with, beneath him, a pyramid of contented and industrious children, grandchildren, and serfs. (Coetzee 1988:6–7)

Note: black "serfs" enter only as an input, as manpower, for an enterprise already designed (cf. Rutherford 2001:85). Taken together, the dream topographies constitute an almost Jeffersonian progression toward yeoman farming: whites converting a howling wilderness into the productive garden of settler nationhood. (White) Man and land would become one as the native slipped into invisibility.

Much later, in black-ruled Zimbabwe, precisely this fixation on the landscape helped white farmers to adapt and flourish. Indeed, the Africanization of politics increasingly limited rural whites to farming. And they farmed very well. In the 1990s, commercial farmers carried out a veritable hydrological revolution of earth dams and large-scale, mechanized irrigation. Needless to say, white men blocked rivers primarily for economic reasons—to irrigate crops—but the cultural side effects became nearly as important. Symbolically, this recarving of the terrain united two otherwise contradictory bases for white claims to land—the two dream topographies. Dams constituted an unparalleled agricultural improvement. They demonstrated the continued efficacy of white land ownership and its associated property lines and labor hierarchies. In principle, however, such celebrations of development ran counter to wilderness discourse. Some whites still claim that their forefathers initially settled on virgin land.[3] The larger number, who admit to having trespassed against African farmers, frequently argue that whites *understand* the virgin bush better. They know it, value it, and preserve it. How can they, then, justify the impounding of streams and rivers? White farmers turned this criticism on its head. Dams and reservoirs, they believed, actually *enhanced* natural waterways. In other words, the same hydrological revolution that industrialized the bush also demonstrated good ecological stewardship. The dream of farming and of wilderness became one.

Dams thus served as multipurpose fetishes of white belonging—an aquatic fix to whites' political dilemmas. For Euro-Africans on holiday, Kariba's shoreline provided aesthetic comfort. In the productive pursuit of agriculture, dams affected whites more powerfully still, rooting them in Zimbabwe's soil. Other settler populations—notably Kenya's whites—limited their financial exposure in-country. After independence, relatively benign inducements gradually shifted them from the highlands, freeing up estates for black businessmen (Rothchild 1973:374; Uusihakala 1999:28). In Zimbabwe, by contrast, whites redoubled their investments in infrastructure even as the black-ruled state repeatedly threatened to remove them. In retrospect, they seem to have lost touch with economic and political constraints. In another sense, however, whites were investing in identity, and dams bore a heavy symbolic load. Whites I met in 2002 and 2003 described themselves as equally farmers and dam-builders. Those who fled persecution to the safety of Harare missed their land and its artificial land forms. This, second set of assets—whose spectacular loss could have worsened their anguish—actually gave them comfort. Thanks to the dams, whites left the commercial farms with their pride intact. Although irrevocably dispossessed, they still felt like the true owners of the highveld. Impounded water, in short, helped hydropower whites' enduring sense of entitlement to land in Zimbabwe.

Geography and Whiteness

Euro-Zimbabweans defy spatial categorization. The first white settlers—an amalgam of Anglophones and Dutch-, French-, and Scottish-descended Afrikaans-speakers—crossed the Limpopo from South Africa in 1890. They soon welcomed immigrants directly from Britain, from Britain by way of Asian colonies, and from southern Europe. In one sense, this plurality of origins made Zimbabwe a "global ethnoscape."[4] Yet, unlike the South Asians to whom Arjun Appadurai applies this term, Zimbabwean whites have refused a global identity. They have consistently struggled to enroot and reterritorialize themselves. In 1923, settlers voted overwhelmingly for self-government—as a colony—rather than for continued administration from London. Nearly two generations later, in 1965, the Rhodesian Front government unilaterally declared independence from Britain. Whites then fought a seven-year war against two guerrilla armies. They lost, but the war itself drew them together. Although many left after independence in 1980, those who stayed considered themselves patriots, rather than expatriates.[5] Among them, some soon demanded the status of a native, a claim that did not tend to stick in wider discourse.[6] Whites—while undeniably cosmopolitan—yearn for a parochial identity.

In part, they have succeeded in giving local meaning to even the most global aspects of their history. Virginia, for instance, lies on arable highveld east of Harare, close to the town of Marondera (Figure 4.1).[7] Settlers did not name their Virginia after the American one—at least not directly. They named it after one of Virginia's crops: tobacco. Columbus and Cortes had originally brought tobacco to Europe from Cuba and Mexico, respectively. In 1585, Sir Walter Raleigh named the original Atlantic Virginia after his virgin queen, Elizabeth. A generation later, another Englishman, John Rolfe, experimented with tobacco in Jamestown, Virginia. Rolfe returned to England in 1616, bringing new varieties and an Algonquian wife. These movements generated the famous American tobacco industry and its "Virginia strain" of light, flue-cured leaf.[8] Cultivated by African slaves, tobacco made white men into Virginia gentlemen. Rhodesian farmers turned to the same crop and African labor for a similar uplift. Yellow, flavored leaves soon became the marker of colonial success—not least in "Virginia," Marondera, and the tobacco belt east of Harare. "Over tens of thousands of then desolate acres," recalled Edward Harben, former vice president of the Rhodesia Tobacco Association, in a co-authored book, "a vegetable El Dorado was . . . brought into being."[9] His veiled references to empty land and Cortes complete the circle of tobacco's history: an Amerindian crop—grown by an English-Indian couple, popularized in Europe, and transplanted to Africa—miraculously justifies whites' position in Zimbabwe. With such aptitude for meanings and materials, surely whites could make their home in both Virginias or anywhere in Africa.

Whites' actual movements in Zimbabwe, however, betray a distinct caution. Alert to the land's environmental unpredictability, whites advanced with trepidation and backward glances to Britain. Nineteenth-century "non-cosmopolitan" theories of climate suggested that whites could not survive the heat of the tropical "torrid zone" (Redfield 2000:192–199; cf. Price 1939:194–204). By 1890, newly documented plateaus gave reason for hope (Ravenstein 1891:35). Altitude could mitigate the effect of latitude. In that same year, the British South Africa Company's "pioneer column" of settlers crossed the Limpopo from South Africa and settled the central highlands of what is now Zimbabwe. White-owned estates soon traced the major watersheds, including the line between the

Save and Mazoe catchments, where Marondera and Virginia lie.[10] By 1901, the company described Rhodesia's upland climate confidently as "as healthy and bracing as can be found anywhere" and promised that "children may grow up there as strong as they would at home [i.e., in Britain]."[11] At 1,500 m above sea level, malaria presented only minimal danger. Against the sun—the one remaining threat—whites armored themselves with pith helmets and umbrellas (Kennedy 1987:110–114). Still, doubts persisted. Over lunch outside Marondera, a farmer confessed, "We [whites] shouldn't be in Africa because we are made differently." The plateau's air was too thin for her: "We haven't got the noses that they [blacks] have."[12] Most whites in Marondera inhaled without complaint. Yet, the lowveld—parts of which were once denoted on maps as "not fit for white man's habitation"—made many whites uncomfortable (Fuller 2001:161; Wolmer 2001:33). White writers still describe the valley's hottest period, October, as "suicide month."[13] If only indirectly, Zimbabwe's environment could still strike a European dead.

Precipitation also has given whites, particularly farmers, ample cause for discomfort. Zimbabwe's rainfall is as intemperate as its heat, differing from that of Britain in both seasonality and intensity. On the highveld, rain falls only from late October to early April. Almost from their arrival, whites have revelled in the long dry season. In 1928, two ex-missionaries founded the Anglican Ruzawi School for whites outside Marondera because, as they later wrote, the area boasted a "climate as nearly perfect as could be found" (Carver and Grinham n.d.:25). Farmers, however, found the climate far from ideal. H. K. Scorrer, who trained Marondera's early settlers in agriculture, had difficulty raising drought-resistant livestock. "If we don't go too fast with European blood [in breeding cattle]," Scorror predicted in 1908, "we shall get a beast that will stand the climate of this country."[14] While aridity hindered animal husbandry, downpours destroyed crops and eroded topsoil. Zimbabwe's rainfall spiked violently and unpredictably: events of 100 mm were not uncommon. In 2001, a 150 mm storm breached the smaller of two dams on Airlie estate: "literally the cloud up above just drops everything that it has," recounted the farmer, still with an air of disbelief.[15] Such conditions—implicitly compared to English mildness—made agriculture an extreme sport. Referring to Zimbabwe's "vindictive climate," Harben and his co-author praised tobacco farmers for "a ruthlessness, an independence, a physical endurance and courage, a coming to terms with harsh forces with which their fellows in more sophisticated societies have long lost contact" (Clements and Harben 1962:188).

If the land and climate challenged rural whites, it also filled them with awe. In the midst of losing their farms, they felt and remembered a sense of wonderment. When I met Steve Pratt, he spoke first of his fears. As the provincial representative of the Commercial Farmers' Union, he was dashing to occupied farms to negotiate for the release of white families and their movable property. Whites, he said, had been "hugely confident" but were now gripped with "a kind of angst about their identity." He was feeling it too. Still, he loved Africa, he said, and experienced "an exhiliration" in the bush. I asked him to be more precise. "When the rain comes," he began, "that smell! When you can hear a storm sort of approaching . . . " As a child on a Marondera farm, he knew the river would rise outside his window in an hour. He recalled listening expectantly. Failing to describe the sensation in his own words, he cited a line from Shakespeare's *The Tempest*: "Show me the magic!"[16] The literary reference was even more apt than Pratt suspected. News of America's Virginia inspired *The Tempest*, a work that—according to Leo Marx (1964:34–36)—presaged the American pastoral ideal of wilderness and agrarianism (quite similar, in fact, to South Africa's pastoral canon). Pratt's imagination and profession combined the same opposites: empty land and efficient farms. Cathy Buckle, who wrote fiction and political literature (see Chapter 2), described Zimbabwe as "so wild just on your doorstep . . . modern but yet not."[17] The landscape defied categorization. For farmers—especially those as imaginative as Buckle and Pratt—the highveld was home without being normal, reliable, or safe. To belong there remained a work in progress.

INTENSIVE CONSERVATION

For farmers less artistic than Pratt and Buckle, collective efforts gave expression to the quest for belonging on the highveld. Chief among these was soil conservation, which had concerned the colonial government for most of its tenure. In the 1930s, the state had encouraged farmers to combat erosion. Edward Alvord, the American-born chief agricultural officer, wished, at all costs, to avoid an African version of Oklahoma's Dust Bowl.[18] Initially, he and his colleagues faced an uphill battle: both farmers' economic survival and the drive to settle the highveld with Europeans overrode concerns about long-term fertility. Simply put, Rhodesian farmers mined the soil without check for at least four decades. In 1941, however, the colony created a Natural Resources Board, which, in turn, fostered intensive conservation associations (ICAs) at roughly district level (Phimister 1989). Organized by farmers themselves, ICAs encouraged agriculturalists to construct and maintain broadbase terraces.[19] The Resources Board delegated to them authority over conservation in black-held land outside the commercial farming areas. In these zones, the ICAs never succeeded. Smallholders working undersized, sloping plots with little labor power could not comply with conservation rules and interpreted them as meddlesome and even racist.[20] Blacks surely noted the fact that most ICAs met in the whites-only social clubs of their districts. By the 1990s, blacks' refusal to cooperate had turned the ICAs exclusively inward. When they mentioned blacks at all, they complained about workers causing erosion on farms. As their main activity, ICAs inspected the members' terraces and dams, keeping records, issuing warnings, and, if all else failed, levying fines. As deeply committed to private property as it was, white society permitted these intrusions. Only land seemed to trigger such acquiescence and cooperation. Labor—which was as scarce as soil—did not generate a single local-level organization in white Zimbabwe. Of course, whites colluded privately over wages, and many complained personally about the "labor problem" to and through the national-level Agricultural Labour Bureau.[21] In formal terms, however, the districts' white bureaucracy largely ignored blacks and even its white members' concerns regarding blacks. Virginia's teamwork was environmental.

The imperative to protect soil followed from settlers' initial decision to occupy the watersheds. Altitude lowered the temperature, making the plateau a more comfortable and salubrious home than the lowlands. The lower temperatures, in turn, allowed moisture to precipitate, bestowing roughly 800–1,100 mm of rainfall on the highveld, as opposed to the lowveld's mere 500 mm. The wetter climate, of course, benefited agriculture, but it came at the cost of a more arable topography. Whereas, along the Zambezi, Save, and Limpopo Rivers, Zimbabwe's lowlands lie flat, at altitude, the country breaks up into granite outcrops, streams, and uplands. Virginia, for instance, straddles the Macheke, Shavanhohwe, Munyuki, and Nyadora Rivers, the last one falling 400 m in 35.5 km (Figure 4.2).[22] Of family-owned farms ranging from 500 to 1,500 ha, farmers

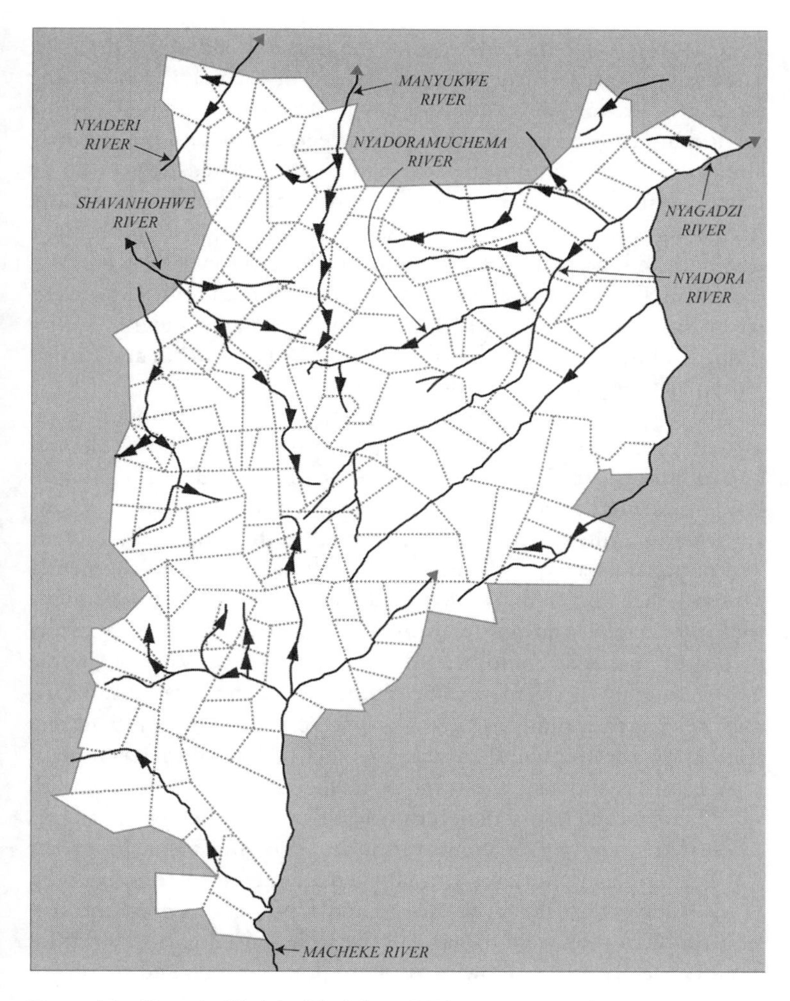

Figure 4.2 Dams in Virginia, Zimbabwe, 2000

Source: Map by Michael Siegel, Rutgers Geography Department.

considered only 100–500 ha flat enough to plant crops. Even on these arable patches, gradients generated ferocious runoff that could destroy the soil profile. To minimize such damage, land owners devised means of "mechanical conservation," specifically, broadbase terraces (known locally as "contours"). Farmers built terraces slightly off the natural contour, at a 1–2 percent slope, and separated by 1 m of elevation. They planted grass on the tops and along the drainage waterways located at the downstream end of each ridge. When the terraces worked, water would run down the slope for no more than a vertical meter, then take an abrupt, 90-degree turn, and move slowly along the terrace, infiltrating the soil to the desired degree. Some farmers elaborated still more intricate systems of holding earth and harvesting water. Perpendicular to the contours, Doug Dunford built tie ridges and, perpendicular to them, "little dams" every 1.5 m. Each dam created "a very large, bath-sized sort of thing to hold water . . . so it can take probably four inches of rain in a night and not spill a drop."[23] Dunford effectively harnessed the 100 mm storm and turned Virginia's topography to his advantage.

Although whites took credit for such ecologically minded farming, it derived as much from pre-existing social and ecological circumstances. Virginia's small community of 72 land-owning households presented ideal social conditions for the ICAs' form of self-organization and self-policing.[24] Although differentiated by income and national ancestry—British, Greek, Afrikaner, and more recently Dutch—they increasingly identified themselves as a unitary white minority. Every farm automatically belonged to the ICA, and any owner or manager could attend the meetings. In Virginia in the 1990s, roughly five farmers came monthly to such gatherings, invariably held at the country club.[25] A respected, conservationist farmer chaired the meetings, and another member (almost always a woman) took minutes and sent the minutes to the entire community. With such institutional transparency, the mere threat of labeling and stigma motivated many a lazy conservationist. Also, the behavior of the soil and terraces themselves virtually demanded cooperation between farms. Once constructed, terraces could rapidly exacerbate the erosion problem they were meant to solve. The raised part of a terrace would develop breaks, allowing water to pour through and run down to the next terrace, possibly breaking that one as well. Especially in the prevalent sandy soil, fields became gullies, known among the farmers by the Shona word *donga*. With an affect bordering on horror, Bruce Gemmill, ex-chair of the Virginia ICA, reported seeing on at least one commercial farm "a *donga* that will drop a London bus into it."[26] Large-scale erosion of this nature could diminish the productivity of an entire watershed. Loosened soil would enter streams and eventually silt up reservoirs used for irrigation. Especially during the dam-building boom of the 1990s, erosion threatened the entire hydrological basis of white wealth. These combined motives of environmentalism and self-preservation gave Virginia's ICA an unparalleled moral authority.

In this context, blacks' indifference to the ICAs confirmed whites' low opinion of them (and whites' high opinion of themselves). The ICAs continuously combated black recalcitrance. Although they excluded peasants from the ICA meetings, the associations invited them to district agricultural field days—for competition and instruction.[27] Black commercial farmers who bought land in Virginia after 1980 *could* attend meetings. Yet, they chose neither to join nor to obey the ICA. Their "problem farms" appear with disproportionate frequency in the minutes of Virginia's association. In 2002–03, many whites dwelled on this discrepancy, describing blacks in general as deficient conservationists. "The communal land boundaries," complained one farmer in 2003, "were like [bare] highways."[28] Such whites felt they carried the conservation burden alone. Said Gemmill, "*We* are the keepers—or were the keepers—of the countryside."[29] He was probably thinking of Dave Stevens, his successor as ICA chair, who was murdered by a death squad in 2000.[30] For Virginia's whites, this killing framed the moral opposition perfectly: a great conservationist—"Mr. Green himself"—political activist, and fluent speaker of Shona against an amoral, destructive state.[31] As recalled in 2002–03, Stevens and the Virginia ICA stood at the pinnacle of collective stewardship. "They are such conservationists, these men," extolled one former member. "Their life is in the land."[32] Conservation had become a discourse of hagiography and nostalgia.

In less politically charged conversations, Virginia farmers often reminisced about a quite different benefit offered by the Virginia ICA—a visual experience. Land owners had already seen their estates from the air. In the 1960s, the government Department of Conservation and Extension (Conex) had used aerial photos to make detailed farm plans—photo mosaics that farmers in 2002–03 still displayed with pride in their living rooms or offices. The ICA gave firsthand access to the aerial perspective. Twice per year, the group rented a light aircraft—often owned by a member—and flew the district.[33] The aerial view revealed all secrets. A broken terrace, said one ICA member, "sticks out like sore thumbs."[34] "Fly over it," explained one farmer with reference to the maize crop, "and you can see immediately that it's not as great as you thought it was."[35] The ICA also detected deforestation, eroding dam spillways, and all manner of changes to the soil and vegetation. Farmers revelled in this panopticon effect—what Gemmill called the "eyeball inspection"—and even considered using satellite and aerial photos.[36] Yet, for all this attention to infractions—and their dutiful recording in the ICA minutes—farmers recalled good behavior much more readily than bad. The flyover "made a huge impression," said one farmer, "all this potential production."[37] Gemmill himself spoke of production with greater specificity: "Fly over, and there were dams everywhere . . . [Virginia was] sparkling with farm dams all over the place."[38] In short, the ICAs gave farmers the ability to see commercial agriculture from above, and they liked what they saw.

More broadly, the ICA and its aerial tours helped promote an aesthetic sensibility—one that drew attention to certain aspects of the land and rendered others invisible. Farmers were used to reducing a landscape to geometry. The Conex air photos traced the boundaries of fields and waterworks in clear lines. Contour maps, which the farmers also used and displayed, similarly represented the relationship between slope and water in linear fashion. Broadbase terraces constituted another set of curves, the less interrupted the better. Farmers took a keen interest in this geometrical, perspectival aspect. On the veranda of his estate, I asked a Marondera farmer what it meant to be a good farmer. I expected an answer related to technique and yields, but my informant dwelled on forms of cleanliness:

> You can see good crops when you drive past . . . [On] a farm that looks well looked after, . . . the fencing is there. The roads are graded . . . You had other farms that looked very untidy . . . [They] didn't give a good impression.[39]

Improvements, in other words, caused a farm to shine—even when they were not ecologically recommended. Removing stumps, for example, destroyed indigenous woodland permanently but left an uninterrupted field. As one farmer opined, coppicing, or regrowth from the stumps, was not only "so ugly" but also typical of blacks' improper land management.[40] Needless to say (among whites), the erosion-battered communal lands were unsightly almost beyond redemption. Black Africans do not appreciate "beauty and nature," asserted one white farmer, but "*We* must live with it."[41] He neglected to mention that the sweat of black Africans had made his farm as beautiful as it was. Indeed, the entire aesthetic sensibility of white farmers tended to render black labor invisible. Virginia farms employed up to 300 workers and housed most of them on the farm. Yet, like California growers and British gentry, owners saw the landscape as a product of whites' culture rather than of blacks' exertion (Mitchell 1996:26; Williams 1973:46). Whites, they implied, had encountered the land and, singlehandedly, made it a sight to see.

This tacit man-land story conjoined production and beauty. Without the effort one might expect, whites reconciled two seemingly distinct principles of land use: landscapes of leisure and working landscapes, or spaces of consumption and spaces of production (Lefebvre 1990; A. Wilson 1991). In Virginia, what was pretty was also frequently useful. Terraces, for instance, beautified the topography while saving topsoil and improving yields. Although economic arguments initially drove whites to install terraces, an *aesthetic* disgust with erosion added to this motivation. Once terraces graced the hillsides, whites enthused about them in unabashedly aesthetic terms. Economically beneficial practices appeared—almost by definition—to be ecologically advantageous *and* beautiful. There were exceptions, of course. In 1991, Gemmill tried to abolish a practice that was of obvious economic merit: using free, indigenous timber, rather than purchased coal, for curing tobacco. Deforestation, he argued at the ICA's annual general meeting, destroyed both ecology and pleasing prospects. As long as "the trees remain," he foretold, "rural appearance and character remain for the benefit of present and future generations."[42] What had he meant by "rural character?" I asked Gemmill at my home in Harare in 2002. "The person like yourself who drives in a motorcar out of town," he explained, "should be able to share in that view . . . [so] that you are happy to go out there."[43] Despite some cutting of trees, Virginia still held enough character to attract a "tourist gaze" (Urry 1990). Despite armed conflict—which had forced Gemmill off his own farm mere months before our meeting—Virginia still grew top-grade tobacco. Ingeniously, whites made a landscape that rewarded the eye and the bank account simultaneously.

HYDROLOGICAL REVOLUTION

If terraces maintained white Virginia's "rural character" in 1991, then, farm dams vastly improved it in the ensuing years. Whites, of course, had blocked waterways in Zimbabwe long before that. In the lowlands, the colonial governments of the Rhodesias and Nyasaland dammed the Zambezi in 1959, creating Lake Kariba (See Chapter 2). Whites recognized such accomplishments as epochal and took full credit for them. "To the air traveler," began a 1969 tourist article (invoking the bird's-eye view):

> Rhodesia's countryside is a panorama spangled with the flashing mirrors of a thousand lakes and dams. From the vast reaches of Lake Kariba to the humblest farm pond, every one of these is a legacy of the ingenuity and enterprise of generations of Rhodesians. Nature formed Rhodesia without lakes: each one of them has been built by the hand of man. (Anonymous 1969: 4)

The "man," needless to say, was white, and after independence, whites began to construct dams and farm ponds that were not so humble. "Everywhere you could catch water, they caught the water," recounted a Virginia man who came to the district in 1989, just in time for the "hydrological revolution" of the 1990s.[44] At the beginning of 1989, there were only seven impoundments in Virginia that held enough water for irrigation. Between then and the end of 1997, Virginia farmers built or raised another 38 dams, enhancing the district's storage capacity by a factor of seven (Figure 4.3).[45] Roughly one in two families engaged in this effort—probably similar to ratios elsewhere in the highveld.[46] Building halted only when, in November 1997, the state designated 1,471 farms nationwide for compulsory acquisition; not a single dam went up in Virginia

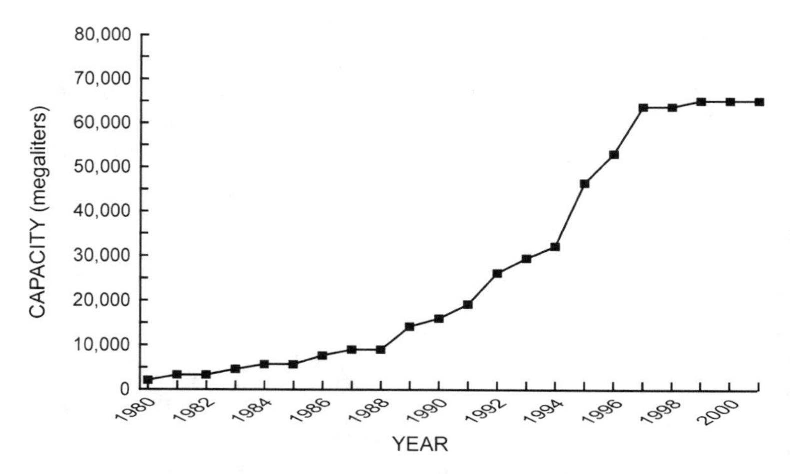

Figure 4.3 Aggregate capacity of reservoirs in Virginia, Zimbabwe

Source: Chart by Michael Siegel, Rutgers Geography Department.

in 1998. Nonetheless, tobacco continued to boom under irrigation, growing two or three crops per year. Growers nearly bankrupted by the horrendous 1991–92 drought weathered subsequent dry spells. All in all, Virginia "underwent a farming transformation," wrote a displaced white in 2003, "from a rather drab farming address into an up-market place to be."[47] These landowners grew rich, and—just as important—they grew entitled. Dams restored whites' sense of ownership and gave them a purpose.

Many whites built dams, in part, so as to secure their ownership of the land. In the 1990s, commercial farmers faced the serious prospect of losing the highveld. In 1990, provisions of the Lancaster House constitution—designed to protect whites politically and economically—expired automatically. Whites had already surrendered their guaranteed parliamentary seats (in 1987). Now, they lost their all-important veto power over land redistribution, the practical consequence of a willing buyer – willing seller format in effect between 1980 and 1990.[48] Suddenly—in a shift of far more legal significance than Zimbabwe's independence—the state wielded the power to confiscate land without recourse and redistribute it to black farmers. At the same time, a strategy to retain their land presented itself. The 1992 Land Acquisition Act, which eventually enabled the designations of 1997, permitted the state to take land without compensation—it having been in theory stolen by the pioneers. Fortunately for whites, the state *would* reimburse landholders for improvements they had made.[49] This loophole revised all economic priorities. Dams, tobacco barns, even workers' housing, which had previously been considered *desirable* under the right conditions, now appeared absolutely *vital* under any conditions. Fortuitously, Structural Adjustment reforms of 1990 allowed farmers to keep a much higher fraction of export earnings. They accumulated capital and could borrow more from the banks. "Guys spent . . . bags of money on improvements," marveled one farmer, referring to an apparently oversized reservoir on the White Gombola River, just outside Virginia.[50] "The more you've invested in your property and the more infrastructure you've got," confided a Marondera farmer, " . . . then they might go and look for a less developed property."[51] Farmers modernized their estates beyond government's price range. The strategy seemed to work: high costs—and, especially after 1997, legal challenges from the farmers—stalled land reform during the 1990s.

Recalling that ten years' grace period, most whites tended to downplay such political calculation and to highlight economic national service. In 1980, they remembered, Mugabe promised whites that they could stay as long as they produced for Zimbabwe. Whites already possessed the requisite personal ambition and entrepreneurial spirit. Explained a Virginia farmer relocated on the outskirts of Harare, "We were a generation or a nation of developers."[52] True capitalists, whites reinvested profit in their farms, rather than stashing all of it in overseas bank accounts—a pattern they identified with Zambian white farmers. "You stagnate; you die," warned Johann Swanepoel, an Afrikaner and one of the few farmers still cultivating in Virginia in 2003.[53] Having so invested in the land—in a fashion that recalled the colonial beneficial occupation clause—commercial farmers felt that they had *earned* a place on the postindependence highveld. And the beneficence of their occupation was patent. Under irrigation, secondary and tertiary crops of tobacco doubled and trebled foreign exchange (forex) earnings—revenue that the state taxed ever more rigorously. Forex proved white farmers indispensability. So did the brute, material infrastructure. Dams, claimed one farmer responsible for one of Virginia's largest impoundments, were "the turnaround of this country."[54] Trusting that they could build and harvest their way to security, farmers seized on any hopeful evidence they could find. In 1995, for instance, Mugabe visited the Virginia Club by helicopter and in the company of ICA member Tom Sweeney. According to Sweeney, the president—gazing downward—remarked to an aide, "Isn't it wonderful the way we built all these dams?"[55] Apparently, Mugabe thought his government had constructed the embankments, but that mistake hardly mattered. Virginia farmers—even if they did not hear or believe Sweeney's story—*expected* the state to appreciate the dams. Surely, they reasoned, those who impounded water to such good effect deserved a reprieve from land reform.

But reservoirs were not natural. In order to rejoice wholeheartedly in the new hydrology, farmers first had to reconcile dams with their self-image as ecological stewards. Surely, each artificial impoundment had damaged the environment, drowning the valley upstream and dessicating it downstream. In 2002–03, Virginians did not deny this damage, but—through various improbable theories—asserted that dams had enriched habitat and hydrology in other ways. An impoundment "is an improvement," insisted Constantine Gavras, who had memorialized his dam on video. "When you've got hundreds of dams in the country . . . you increase your rainfall."[56] He was referring to the effect of added evaporation on highveld microclimates—a relationship that has never been measured and probably does not exist.[57] More plausibly, Virginia growers claimed to have improved the flow of the Nyadora and other rivers. In 1988, another farmer blocked the Chikumbakwe, a tributary of the Nyadora that ran only in the rainy season. Due to seepage through this and other earth structures, he told me, "rivers run all year round."[58] Even if only a trickle ran through and dried up, the next dam downstream would revive the stream. "The more dams on a river the better," concluded Henk Jelsma, adding saltily that when his pre-impoundment river ran dry, "I couldn't hardly have a crap myself without flushing it [by hand]. It was desperate!"[59] Clearly, Jelsma and his river benefited from the dam in multiple ways. Indeed, because seepage varies directly with the square of the height of a porous dam,[60] the higher dam walls of the 1990s raised dry season flows exponentially. Of course, the newly perennial stream may drown plants and animals adapted to annual dessication. My informants did not appear to be aware of this complication, a consequence of the artificial nature of Virginia's new lakes. The aquatic mania seemed to blind them to all negative effects of water—except, of course, erosion.

Actually, dams could easily cause erosion, and this risk brought them to the attention of the Virginia ICA. As with terraces, the ICA used its monitoring role to pronounce on good and bad stewardship. In this case, it directed criticism not against blacks—for they did not possess dams—but against mostly white engineers and builders. The problem centered on spillways and return channels. Engineers designed impoundments to pass water in the rainy season. Don Lanclos—a former Conex officer who had planned many of Virginia's dams—looked for rock close to the surface so that spilling water would carve a hard return channel to the riverbed.[61] It was precisely this practice to which the ICA objected. Soil removed from return channels, as they eroded to rock, eventually clogged pools and killed aquatic life farther downstream. "The issue must be pursued," record the minutes of a 1996 ICA meeting, "because of the mess being made on our rivers."[62] The following year, a dam under construction wrought much worse havoc. The ICA chair reported somberly to his association, "Some 20 km of complex riverine ecosystem below the [Royal Visit] dam was scoured away and the riverbed now resembles a lifeless moonscape of rocks and sand."[63] Contractors, it seemed, had fallen fatally behind schedule. When, on rare occasions, the farmer himself bore responsibility, the ICA put matters delicately: "WET!!!" Gemmill alerted a meeting at the height of the 1998–99 rains, "Whaley dam in serious trouble—spillway problem. Erosion has been huge . . . Problem seem sot [*sic*: seems not] to be the engineer[']s fault—wrong site."[64] At another level, Whaley and all farmers were obviously liable for dam-induced erosion. *They* had decided to block Virginia's rivers. In 2002–03, none accepted this ultimate responsibility. Packing for New Zealand, the owner of Royal Visit blamed the contractors and then showed me his photo album of the dam's construction, collapse, and reconstruction.[65] Water, even when it caused an erosive disaster, could still fill whites with pride.

Having built dams, farmers were obliged to reorganize their terraces. Typically fields lay on the slopes surrounding a low-set reservoir. Therefore, Zimbabwean commercial farmers had to pump water uphill.[66] Fighting gravity in this way required elaborate technology and imposed material constraints. First, farmers had to install electric or diesel-powered pumps. Second, since canals would not hold water moving uphill, irrigators also had to lay elaborate networks of underground and above-ground movable pipes. Zimbabwean manufacturers made such aluminum pipes only in 9 m segments and only in straight or right-angle pieces. Suddenly, the curvilinear pattern of contour-hugging terraces made no sense. To use the equipment of irrigation, farmers would have to redo their terraces in a rectilinear fashion. This "squaring up" of fields occurred in Virginia over the 1990s, transforming arable land into strips 9 m wide and multiples of 9 m long. This grid differed aesthetically from the intricacy of Kariba's shoreline, and farmers chiefly appreciated it for its managerial, rather than aesthetic, qualities. "Parallel layouts" simplified relations between the farmer and his labor force. In the past, farmers and foremen had allocated piecework according to field areas, but no one had measured the areas between terraces with precision. Hence, farmers judged them by sight. The resultant ambiguity led to delays and disputes with employees. (cf. Rubert 1998:178; Rutherford 2001:110–111). Layouts, however, brought Taylorist, Fordist techniques to rural Zimbabwe. "It was a work efficiency scenario," explained Les Wood, the former water coordinator for Virginia.[67] Within the grid, "it's easy to calibrate" piecework, enthused Johann Swanepoel, "now you don't always have to stand at his [the worker's] back."[68] In other words, Swanepoel's topographical designs—shown to me on vellum sheets—replaced face-to-face contact. Layouts gave the clearest material form to that unmediated (white) man-land relationship so valued in highveld culture.

Layouts also problematized that relationship by raising the specter of erosion. The curvilinear form of terraces had allowed them to hold to a shallow 1/250 slope, keeping water at low, safe velocity. Once straightened and made parallel, however, waterways inevitably cross-grained the landscape (Elwell n.d.:7). If farmers wished to maintain the 1/250 gradient, they would have to close off layout segments where the land dipped. Understandibly, farmers were loath to take precious arable soil out of production, and many were tempted to extend layouts until they created dangerously steep gradients. Such a practice courted erosive disaster, and the ICA issued warning after warning. In the gentlest tone, Dave Stevens informed the 1993 annual general meeting, "Because of the nature of our farms, we cannot all have parallel contour systems."[69] Three years later, Stevens spoke more explicitly and with climatological detail: "Members are urged to review their land layouts very carefully and to provide a sufficient area of waterway beside and within lands to cope, not just with moderate rainfall, but also with those 4 inch storms."[70] Yet, the problem persisted. In 2002–03, Virginia farmers recalled layouts tilted recklessly at 1/60 gradients. Such farmers, many of whom were then abandoning their estates, were criticized in absentia. "Your priority is to look after the land, not to make your life easier," chided one farmer in an interview.[71] Layouts, recommended another whose ridges ran at 1/300, worked only "if the lie of the land is suitable."[72] Obsessed with topography, the more conscientious farmers relearned and recommitted themselves to the broken landscape of the highveld.

At the same time, and in a somewhat contradictory fashion, conservationists grappled with the new aesthetic possibilities of layouts. "Squaring up" straightened the curvilinear format characteristic of broadbase terraces—to the delight of many farmers. Indeed, the grid almost became a goal, in and of itself, related to but distinct from the economic advantages of irrigation. Gemmill, while ever-vigilant against badly made layouts, thrilled at the sight of well-made ones. "We could pick that up from the air," he reminisced, "a beautiful grid." Indeed, Gemmill had converted some fields to rectangles even before the installation of his irrigation dam in 1991. "I did it for easier layout," he confessed, "it all seemed tidy to me." Gemmill appeared to recognize where this fastidiouness could lead. Symmetry threatened to supercede conservation. Rather than round off a corner—to allow for some topographical or ecological obstacle—farmers would run pipes and ridges straight through it. "Don't bulldoze out trees where you don't need them," he advised me in the same conversation, "just because you want a straight edge to your land."[73] Les Wood, also an upstanding conservationist, seemed entranced with such geometry: "Something that looks squared and laid out and done properly has a certain appeal. Doesn't it? . . . [It's] aesthetically pleasing . . . As a people, the whites, generally speaking, like straight lines." Given his and his co-ethnics' preference for grids, Wood advised farmers on a minimal form of layouts. Rather than extending a rectangle into dubious areas, he suggested foreshortening it dramatically. "Pull back, take it out," he exhorted. Farmers who followed his advice sacrificed sizeable chunks of perfectly arable land. Wood suggested that such marginal land did not produce high-grade tobacco in any case.[74] Still, many farmers would surely have seen his solution as economically suboptimal—but implemented it anyway. Conservationist aesthetics, drawn on vellum, set the course for many a tractor in the 1990s.

Virginia's hydrological revolution, in fact, conjoined beauty, production, and belonging even more thoroughly than had the earlier terraces. Swanepoel, who in our first conversation had explained the efficiency of labor, later summed up his entire enterprise in loftier terms. "The obvious thing," he declared, "is to develop and to beautify."[75] This combination of seemingly opposed values did not initially ring true to me. A month later, I asked Tom Sweeney which was *really* more important, economic development or beauty? Of course, dams brought economic benefits, he admitted: before them, southern Marondera had been "a bum-farming area . . . almost a peasant area. I'm talking on a white scale." At root, though, economics and aesthetics were equivalent. "If you have farmed in a series of droughts," he explained to me (an obvious urbanite), "then water becomes a very . . . beautiful thing to see . . . like jewels" when viewed from the air.[76] From their planes, farmers gazed down on the landscape they made and that made so much of value to them. Dam builders found a way to transform the highveld, love its landscape, and belong in Zimbabwe all at the same time. And almost as soon as they grasped it, they lost it. Moving into a gated community outside Harare, an ex-Virginian predicted that whites might one day regain farms somewhere, but "we will never develop them, beautify them as we did. [Rather than invest in them] we will get U.S. bucks outside the country."[77]

A ROOM WITH A VIEW

Dams did not merely irrigate crops. Their nonagricultural attributes—particularly shoreline—contributed a substantial increment of aesthetic value. Recall Zimbabwe's hydrological deficit vis-à-vis Europe: the country contains not a single natural lake and no coastline. Many Zimbabwean whites have felt this lack keenly. They desired water not only to nourish their tobacco plants but, more emotively, to look at while relaxing and smoking tobacco. Before irrigation dams, they could only gain access to Lessing's "long, grey sea" at Kariba and by dint of the most advanced engineering. The hydrological revolution of the 1990s brought more modest lagoons to the very doorsteps of commercial farmers. As bulldozers did the work of small glaciers, reservoirs inundated highveld valleys and, along upland contours, created numerous vantage points from which to view the resulting reservoirs. In Virginia alone, dam construction between 1990 and 1997 increased the district's shoreline from 38 to 203 km (Figure 4.4).[78] Whites found that interface between land and water beautiful, and, still in 2002–03, identified it with European heritage. "I think water has always been a calming effect," said Swanepoel. "We [whites] in Africa always like a nice view and trees, and we like nature."[79]

Some shorelines excelled in providing such sheer, nonproductive beauty. I asked Les Wood which, of Virginia's 203 km of littoral,

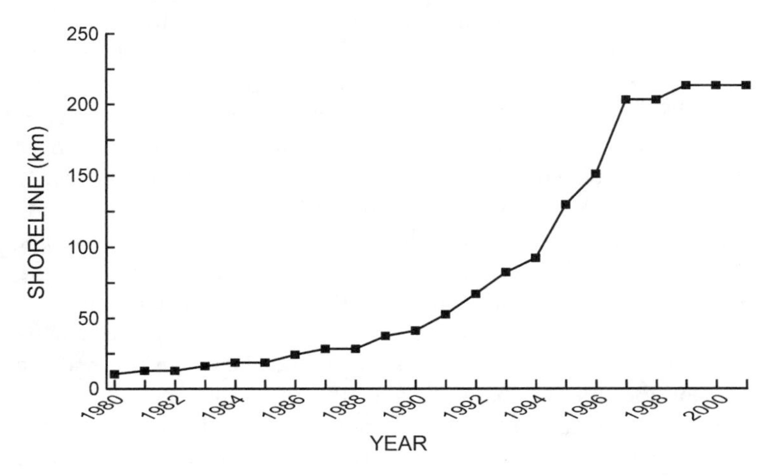

Figure 4.4 Aggregate shoreline of reservoirs in Virginia, Zimbabwe

Source: Chart by Michael Siegel, Rutgers Geography Department.

gave the greatest aesthetic pleasure. He pointed me toward Chingezi reservoir, lying across the Nyadoramuchena river and one of the district's largest by capacity.[80] The owner of Chingezi, Henk Jelsma, had built the dam in 1993 and raised it in 1996. He did not, according to Wood, select the best site for irrigation. The impounded water filled a bowl, requiring Jelsma to pump it up steep slopes to his flatter, arable lands.[81] Yet, in aesthetic terms, Jelsma made an inspired choice. The very same slopes pinched the reservoir basin, forcing water up the Nyadoramuchena and into three tributary streams. The resulting shoreline extended over 9.9 km, the fourth longest in the district.[82] More importantly, the topography above the waterline—creased by four watercourses—created a sense of privacy along the shoreline. People could watch the water without, themselves, being watched. This seclusion, combined with one's distance from cultivated fields, gave Chingezi an air of wilderness. Jelsma himself showed me the littoral. "You've got trees all the way round," he narrated as we walked. "It's very quiet in the bush—virgin, scenic." "Virgin" meant "scenic" and scenery depended on water and on the lines of sight around it. The impoundment of the river—upsetting to another kind of of nature lover—only enhanced the valley's pristine quality. "Idyllic," Jelsma pronounced.[83] He and other whites had once again mastered the highveld's broken topography. What terraces achieved for cultivation, reservoirs accomplished for contemplation.

Still, not every shoreline possessed Chingezi's baroque curves. Farmers with bland littorals could retrofit them for complexity. For Gemmill, it was important to "end with something that wasn't . . . offensive when you walked through there." He and the ICA advised farmers on various ways of "creating a pleasing appearance."[84] Farmers added peninsulas and islands. No one east of Harare knew more about the efficacy of such measures than John Tessmer.[85] A teacher of ecology and manager of his school's private woodland, Tessmer manufactured bird habitats. Although he advised farmers in Virginia—and had even spoken formally at an ICA annual general meeting[86]—Tessmer's greatest work lay just outside the district. On Shiri Farm, Tessmer and the owner had added 260 percent to the length of the main reservoir's shoreline (Figure 4.5). "A duck will only occupy one bay," Tessmer informed told me, and so he designed 12 small bays on Shiri. Better to display the birds, Tessmer constructed walkways into the reservoir. He used anthills to make islands. Finally and most ingeniously, Tessmer scooped out a set of six depressions in the reservoir's bottom that would hold water as

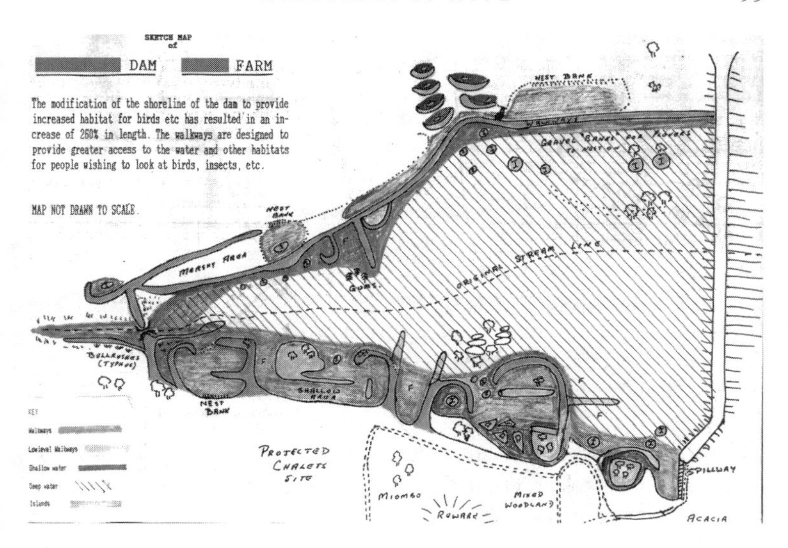

Figure 4.5 Sketch map of Shiri Dam, by P.J. Ginn, ca. 1995

it receded (at the top of Figure 4.5). A large draw-down for irrigation would actually *enhance* ornithological diversity. It worked—or at least observers thought it did. "We pulled the migratory route of ducks over this area," Lanclos boasted. While flying from the Mediterranean to South Africa, he elaborated, the Egyptian goose and knobnose duck actually veer slightly eastward to visit Virginia's reservoirs.[87] Bird counts did not confirm this global ornithological effect,[88] but the symbolism of the assertion mattered far more: birds voted with their wings. After viewing all of black-ruled Africa from the air, they favored Zimbabwe's white highlands.

Pro-avian enhancements to the shoreline benefited underwater species as well. A long shoreline and intricate topography provided habitat—known as "structure"—for aquatic plants, fish, and ultimately for their predators. Occupying the top of the food chain, sport fisherman strove to enhance the biological productivity of their dams. The organization Zimbabwe Bassmasters and, especially, its Virginia–Headlands chapter stepped forward to help them. As the head of that chapter, Graham Murdock, explained, "I am a bass fisherman who looks to create more places to go fishing . . . It doesn't come naturally. You've actually got to create that environment."[89] In fact, one had to create everything about it: the bass—fierce, fighting fish—were imported from the United States and introduced, by

Bassmasters, to new reservoirs throughout Virginia. In the reservoirs, Bassmasters encouraged farmers to dump tires, logs, and other bits of artificial structure. Finally, and most heroically, Murdock actually rescued fish from reservoirs as they evaporated in the 1991–92 drought and transferred them to safe storage. Why did he and other Bassmasters and such a large portion of Virginia's farmers go to such extremes? They enjoyed angling, of course, but it also animated their community. "A lot of these guys like their fishing," explained Swanepoel. "It's social. They go out on a boat and sit there and have their *braai* [barbeque]. It's different from having a *braai* in the garden."[90] The difference lay in the water. Engineered hydrology fit hand in glove with rural white society.

That hydrology could appeal to urban whites as well. In 2000, Virginia farmers began to market the beauty of their water to tourists.[91] In that year, Frank Richards constructed three chalets along his reservoir.[92] Blocking the Nyadora River since 1995, his impoundment boasted Virginia's second-longest shoreline (15.15 km) (Wood 2003). As a further aesthetic virtue, wildlife abounded on Richards's farm. He saw kudu, sable, duiker, and klipspringer regularly and hyena, leopard, reedbuck, greysbok, steenbuck, wild pig, and jackal less frequently. Of course, the same animals roamed widely in Virginia. In this patchy, discontinuous habitat, they used the areas of farms too steep or rocky for cultivation. In effect, Richards found yet another way to use the highveld's broken topography. Among his neighbors, the idea caught on. In 2002–03, I found another five Virginia farmers who had considered chalets. Two of them planned to join their properties as a conservancy and—not satisfied with the existing biodiversity—to stock their land with impala, nyala, and zebra. Still, shoreline was the main attraction, and the conservancy's chalets would have abutted it. "If you're looking at water," explained one of the owners, "and it's pleasant, it's quiet. What a way to relax."[93] More intricate shorelines heightened this sense of calm. Jelsma planned to install lodging in the estuaries of the streams feeding his unusually dendritic reservoir. Guests would enjoy an unobstructed view of the water while ridges obstructed views of fellow guests. This type of optical geometry, Lanclos explained over lunch with Jelsma and me, allowed chalets to "give [guests] the feeling of being completely by themselves."[94] Tourists and their hosts craved human isolation and faunal company—a combination that they called "virgin bush." An ironic, anthropogenic nature was starting to flourish on the highveld.[95]

Almost immediately, however, it was cut short and reduced to a mere rhetorical device. By 2002–03, paramilitary violence had at least deferred the dream of ecotourism (Hughes 2001). Whites cited not extant chalets but the *idea* of chalets as evidence of their ecological stewardship. Black farmers, they implied, never would have aspired to ecotourism. "The world consists of two types of people," Sweeney explained to me, "creators and users." Clearly, most blacks fell in the latter category. Sweeney explained their instrumentalist approach to reservoirs and fish. In impounded water, they sought only food, as opposed to whites' "bottom-line of generating something good and beautiful and valuable."[96] Bassmasters, who practiced catch-and-release, expressed outrage at what they saw as a pervasive black tendency to overfish and even to vacuum reservoirs with nets. When black settlers moved onto his farm, Richards initially established a reasonably amicable relationship. "What killed it [however] was . . . the total destruction of animals."[97] What was being destroyed? Even if tourists had come in droves in 2002, Richards would not have earned much revenue: the chalets charged only the equivalent of US$1 per person per night. Tobacco, of course, did put money in the bank in peaceful times. Yet, in our conversation, Richards did not dwell upon those costs of the occupations. He and other Virginia farmers lamented ecological more than economic loss. Fuming in Harare, Gemmill denounced Mugabe: "he is an environmental pagan, this man. He doesn't give a damn about any aspect of the environment."[98] Some black settlers were even cutting impoundments, practicing gravity-fed irrigation on the downstream side and raising the specter of widespread dam failure. The environment whites had engineered threatened to implode around them.

* * *

Before they were dispossessed, Virginia's whites created what Leo Marx (1964:23) calls a "middle landscape." Like Jefferson of the American Virginia, they imagined a garden, compromising between nature and civilization, between the primeval and the technological. Then, for reasons having little to do with aesthetics, commercial farmers *made* their Arcadian gardens and their geometry of beauty—a success that is all the more striking given the initial conditions. Under a climate of intense storms, Virginia's hydrology and soils behaved—to use Mike Davis's (1998:14) term for Southern California—like "Walden Pond on LSD." Primitive, wild nature raged just outside the kitchen door. Whites could not change the rain, but they changed the texture of the land. Terraces slowed runoff to a stately pace and held soil to soil. If managed properly, Virginia's gardens did not erode. In the 1990s, whites forged another compromise between the highveld's untamed topography and modern technology: the irrigation dam. They bulldozed earth, blocked rivers, and pumped water to agro-industrial fields over which tractors and combine harvesters rolled. Yet, amid the whirring of machines lay a recessed, still space: the reservoir itself, flanked with trees and wildlife. Walden-like, these bodies of water invited transcendence. They also invited political discourse; for the middle landscape is "as attractive for what it excludes as for what it contains" (Marx 1964:138). Most tobacco plantations assigned one role and one role alone for blacks: manual labor.[99] Other categories of blacks, such as peasants and newly minted commercial farmers, could not straddle the divide between primitivism and modernity. After 2000, the new settlers violated nearly all pre-existing codes. They killed wild animals, felled trees, or—as Cathy Buckle wrote after her farm was occupied—"rape[d] the land" (Buckle 2001:10). Portrayed as nearly atavistic in their proclivity toward erosion, blacks did not qualify for admission to the middle landscape. Whites demarcated and regulated their own cultural reserve.

In so doing, whites solidified the man-land relationship vital to their sense of belonging in Africa—and updated that trope for the era of black rule. Those who remained in Zimbabwe into the 1990s identified themselves as liberal in their dealings with workers and other blacks. They would have concurred with Lessing's forward-thinking critique of the fictional farmer Charlie Slatter, who believed "that one should buy a sjambok before a plough or a harrow" (Lessing 1950:13). Yet, having relinquished that infamous hippo-hide whip, most white farmers did not replace it with another instrument or technique that reached across the color bar. Very few of the Virginia farmers I met had ever shared a meal with a black and few intended to do so. They kept away from more mixed venues and social circles in Harare. Intermarriage was unthinkable. In short, rural whites adapted to postcolonialism by withdrawing from, rather than integrating with, the broader nation.[100] Their liberalism engaged with the environment almost as an alternative to society. In place of Charlie Slatter, many Virginians would identify with Alexandra Fuller's (2004:56) white Zimbabwean recluse: "Like the African earth itself, he seemed organic and supernatural at the same time . . . Seeing him on his farm, I couldn't decide if the man had shaped the land or the other way around." Perhaps the land-shaping hydrological revolution substituted for a sociological one.

In this sense, the hydrological revolution was supremely conservative. Virginia's farmers sorely wanted, first, to keep their individual estates and, second, to legitimate their collective status as a land-holding minority. Investing in the highveld advanced whites toward the former goal in a straightforward, Lockean fashion. Each impoundment deepened their sense of entitlement to the estates they owned. This infrastructure also added to the potential expense of nationalization and compensation, making such an event that much less likely. At one level, then, farmers carried out a revolution in hydrology with the implicit aim of forestalling one in property. At another—even less conscious—level, hydrological enhancements could help farmers regain some of the political footing they had lost at independence. Black rule cast the highveld in quite an unfavorable light: as an unjust anachronism, where European-derived people still possessed large swathes of extra-European territory. Could the ecology and beauty of shorelines naturalize such an exotic—even retrograde—sociology? Yes, whites felt in their bones. Dams not only legitimated their discredited minority but admitted it into the moral center of Zimbabwe. Mugabe himself appreciated the impoundments—or so Sweeney had overheard. Amid dams and reservoirs, he, Gemmill, and Stevens fit in. "A white African," said one farmer, "Dave Stevens was it." Surely, *black* Africans would come to recognize this identity. Yet, my informant undercut this praise with a crucial qualifier: "if ever there was a white African."[101]

CHAPTER 5

PLAYING THE GAME

After independence, white farmers accepted an implicit and somewhat incoherent offer from the Zimbabwean state—one quite favorable to themselves. Whites had foreseen retribution, rather than reconciliation, and almost two-thirds of the total population emigrated in the 1980s.[1] Their fears were not misplaced. After all, whites had supported the old regime—the Rhodesian Front and its loyal, legal opposition—almost to a person.[2] This conservative consensus distinguished Rhodesia from its neighbor across the Limpopo, where many whites resisted the state and suffered violent retribution. As compared with the clearly pluralist African National Congress, Rhodesian whites played almost no role in Zimbabwe's guerrilla groups.[3] There was no equivalent to South Africa's End Conscription Campaign either: disaffected draftees merely emigrated. Meanwhile, men of all ages responded sheeplike to increasingly frequent military call-ups—only, in some cases, repudiating their military service postbellum. Bruce Moore-King's war memoir, for instance, blamed "the Elders" for deceiving his "scarred generation." "Call yourself Zimbabwean, with all that word means," he urged his co-ethnics, "and set about rebuilding our war-ravaged homeland" (Moore-King 1988:132). His exhortation coincided with one part of the state's offer: whites would work for the nation. White farmers, in particular, would produce food and export crops to fuel development (Passaportis 2000:101). The Lancaster House constitution also protected commercial farms against compulsory acquisition until 1990. Many anticipated that the government would expropriate farmers anyway—or do so summarily in 1990—but it continued to practice forbearance. Mugabe conveyed a second requirement more subtly: whites must avoid undermining black rule and, specifically, the ruling party. Although the Lancaster House constitution guaranteed whites 20 (of 100) seats in Parliament, their Conservative Alliance party kept its head down and disbanded in 1987 with the expiration of that constitutional clause.[4] In sum, the exchange of economic protection for political silence gave whites more than they had reason to expect.[5]

Even more opportunely, this bargain responded to whites' deeper longings and insecurities. Mugabe relieved the European-descended elite of its "white man's burden"—the obligation to educate, civilize, and lead Africans. Burdensome indeed, that endeavor had foisted an administrative project on whites, forcing many to engage with Africans much of the time. This civilizing mission had also made minority status and foreignness tangible. Precisely to escape from these unpleasant realities, white writers had pursued their imaginative project of ecological belonging. Now, Mugabe licensed all whites to do the same. "For the first time, we were enjoying the country without a conscience," recalled Peter Godwin. "We were no longer in charge and frankly it was a relief" (Godwin 1996:328–329; cf. Selby 2006:122). Political withdrawal, in other words, soothed nerves long jangled by wider moral responsibilities—not least, those of Godwin himself, who, before becoming a journalist, had evaded his second military call-up. Many whites "enjoyed the country" by fishing its waters and ogling its game. Compared with prior wartime austerity, independence restored colonial leisure and privilege. But whites also worked. Farmers doubled and trebled their yields. By the 1990s, they were reshaping the highveld to store water and irrigate some of the best tobacco in the world. Meanwhile, dams, as engines for development, reduced land reform to a distant rhetoric. Commercial farmers became—as Robin Palmer observed somewhat ruefully—"almost a protected species" (Palmer 1990:167). Still, many of them grumbled nostalgically about the general slide in standards since 1980—especially with respect to conservation.[6] Through the mid-1990s, most farmers channeled their environmentalism into private organizations or personal land management rather than into political dissent. Indeed, through their concern for soil, plants, and animals, whites grew increasingly distant from the state. The savannah suggested to whites a world larger and older than government—and one in which they could participate fully. At independence, black society pushed willing whites into the open embrace of nature.

This cozy realignment—based on an ambiguous deal never articulated—could not last. Farmers' economic service brought them gradually into the political realm—more often as whistle-blowers than as cheerleaders. Many started providing seedlings, tillage, and training to smallholders in adjoining communal lands. In some cases, this relationship blossomed into a bid for local government office (in rural district councils). Once elected, commercial farmers fought against the corruption that gradually tainted the ruling ZANU-PF party in the 1990s.[7] Then, in 1998, the party allowed farm workers to vote in local elections, thereby ousting almost all the white councilors. At almost the same time, in November 1997, the Ministry of Lands published a list of 1,471 farms scheduled for compulsory acquisition. Regarding these seizures, owners sued immediately and stalled land transfers for another three years. By then, the Mugabe regime was attempting to revise the constitution and weaken its protections for private property. In response, some whites joined a lobby group known as the National Constitutional Assembly. Stunningly successful, that coalition helped defeat the party's proposal in an early-2000 referendum. Many of the same individuals poured time and money into the Movement for Democratic Change (MDC), contributing to its near victory in the parliamentary elections of June 2000.[8] White liberals reveled in this unprecedented efficacy and collaboration with blacks (Johnson 2000:23). For a shining moment, many felt part and parcel of Zimbabwe. But they had overstepped their bounds. "We had broken the unspoken ethnic contract," recalls Godwin. "We had tried to act like citizens, instead of expatriates, here on sufferance" (Godwin 2006:59). The state responded swiftly. Ceding to demands from its radical wing—and playing the "race card" for elections—ZANU-PF authorized and assisted veterans of the guerrilla war to commandeer white-owned estates.[9] Beginning in 2000, these farm invasions marked the end of a racial reconciliation in agriculture that was neither tenable nor honest.

Jolted by violence and hatred, whites reassessed their position on the highveld and in Africa as a whole. Such discussions took on particular force and direction in Virginia. There, paramilitaries killed the first white, David Stevens, a vocal MDC supporter described by his neighbors as a "white African." They subjected all the farmers to *jambanja*, sustained harassment that confined whites indoors for weeks at a stretch. Threats pushed uncomfortable questions to the fore. "Do we whites hold rights in Africa?" farmers asked themselves, each other, and occasionally me as well. Those who insisted upon the human and property rights of whites and all other citizens lost their farms in 2000. The remainder, still on the land in 2002 and 2003, considered radical compromises. To persist on the highveld, they would largely have to cease farming it and find some other role amid the blacks now using those hectares. At a level that was barely conscious, Virginians experimented with three alternative callings: conservation, evangelism, and—in the spirit of colonial native commissioners—agricultural development. None of these roles placated the state. Still, the war veterans spared an unusually large proportion of whites in Virginia. In comparison to districts completely vacated, 11 of the 75 original Virginia families remained resident and actively farming in 2007. Large enough to set a political example, this group cut deals with ZANU-PF. As some put it, they learned to "play the game." Virginia whites, in other words, honed the skill of operating within the rules set by powerful blacks. They did not like blacks or the black state any the more, but at last and on the least favorable terms, these whites engaged with African society. Implicitly, they struck a new bargain, one surely more honest and perhaps more tenable than that of 1980.

Becoming Political

Whites retreated from two political debates after independence: one regarding land reform and the other regarding the privileges of whiteness in general. On the first of these issues—the seemingly more tangible and contentious one—whites achieved greater success. Nationalist rhetoric had framed land as Zimbabwe's preeminent political and ethical issue (Moyana 1984:127ff). Yet, in the 1990s, commercial farmers removed the highveld from contention through a process Li (2007) terms "rendering technical." The state, for instance, continually accused farmers of hoarding land unethically and unproductively. The Commercial Farmers' Union (CFU), acting as white landowners' representative, countered with anodyne ecological and economic detail. Among these rationales, the CFU cited the paucity of arable land on even the largest farms. Environmentally prudent farmers would not cultivate slopes or wetlands. Furthermore, as economists suggested, the redistribution of large estates to small or mediumholders would result in unviable, postage stamp-sized plots. Dams and irrigation—whose operation required skilled managers—only added to the technical obstacles. By the mid-1990s, such deep complications had detained land reform on a treadmill of increasingly detailed research.[10] When it came to action, the state seemed either uninterested or unwilling to promote large-scale resettlement (Sachikonye 2005:7). While still nominally socialist, the government had adopted a structural adjustment program that discouraged state-led investments. It was no longer buying land—although the budget did accommodate the deployment of troops to Central Africa. Given this standstill on land reform, donors and NGOs were designing projects to make peasants in the communal lands more productive in situ (Hughes 2005:164). In sum, this apparatus of development experts, civil servants, academics, and lobbyists functioned as an "anti-politics machine," dampening outrage and thwarting redress vis-à-vis inequality.[11] Perhaps such institutions and techniques could have shielded commercial farms from political intervention beyond 2000. But, the politics of whiteness itself burst forth and engulfed the highveld.

Debates on race and citizenship had unavoidably trailed whites in the postcolonial period. In principle, whites belonged to the nation like anyone else. Indeed, the army's occupation of Matabeleland—in search of a small number of armed dissidents—shifted attention from the black-white cleavage. In what many interpreted as a tribal conflict, the notorious Fifth Brigade killed at least 3,000, most of them Ndebele speakers (Catholic Commission for Justice and Peace in Zimbabwe 1997:157). In 1987, the Unity Accord ended the violence and ushered in a de facto one-party state. Over the next decade, ZANU-PF gradually returned public discourse to the question of pigment, through the euphemism of "indigeneity." Whites could not claim indigenous status, and, in professional advancement, the state encouraged them to cede ground to emergent, black counterparts. Meanwhile, in more formal terms Zimbabwe perpetuated Rhodesia's policy of *jus soli*, or citizenship based on birth within the national territory (Herbst 2000:240). Many European countries followed a policy of *jus sanguinis*, wherein nationality depended upon ancestry. This disjuncture afforded whites a "flexible citizenship" (Ong 1999): the native-born held Zimbabwean passports and, through ancestry and/or marriage, could acquire foreign documents as well.[12] Then in 2001, ZANU-PF shifted, somewhat underhandedly, toward a *jus sanguinis* principle. Seeking to disenfranchise the increasingly fractious farm workers, new legislation made citizenship contingent on one's renunciation of all other potential nationalities (Muzondidya 2007:135). As expected, laborers with Mozambican or Malawian fathers failed to negotiate such procedures. They simply lost the right to vote. So did many whites, who, even when they did renounce other citizenships, were removed preemptively from the voters rolls. Although transparently corrupt, this disenfranchisement gained some public legitimacy from a material inequality between whites and blacks. Whites could go elsewhere. Even those who clung exclusively to Zimbabwean nationality could appeal in desperation to the embassy of their grandparents. At root, European birthrights and passports followed European descent. "It was nobody's fault," writes Christina Lamb, quoting a farm worker. "That's just the way things were. Whites might lose their farms but they got on a plane to start a new life some other place while blacks lay down and tried to survive on wild fruit" (Lamb 2006:256). Euro-Africans, whose mobility virtually ran in the blood, would never pass as ordinary Zimbabweans.[13]

This distinction did not entirely disqualify whites from politics. In the 1980s, a handful of whites served as ministers or deputy ministers.[14] Even when such national-level opportunities gradually dried up, whites still occupied local positions. Stevens himself stood for rural district council elections in 1993—at the invitation of ZANU-PF—and won. At that time, he belonged to the party and, as a neighbor recollected, "was a strong supporter of Robert Mugabe."[15] Perhaps this sentiment led him to build on his Arizona farm the best workers' housing in the district—a gesture that unnerved some of Stevens's peers (Staunton 2005:466). As an environmentalist, Stevens garnered unequivocal praise. "Mr. Green himself" farmed organically, recalled a prominent ICA member.[16] "To walk around a land with Dave," said another admirer, "was an absolute educational experience."[17] Stevens, then, seemed to embody the sociological and topographical combination so extraordinary among whites—and, apparently, so pleasing to the state. In 1999, however, Stevens's relationship with the party soured. He had uncovered corruption in the Council and appeared ready to blow the whistle. He had also joined the Movement for Democratic Change as a local coordinator. "Let's give ZANU(PF) some competition and try to make a better future for the country," he encouraged his neighbor, Dawn Harper.[18] In April 2000, Stevens hosted an MDC rally on Arizona farm. Shortly thereafter, a death squad took him. The assassins also severely assaulted five farmers who had come to Stevens's rescue and sent Virginia's entire white community into temporary evacuation.[19] Later, my informants eulogized Stevens but also noted his risky behavior. "Maybe, if Dave had stayed out of politics, they wouldn't have killed him," wrote Harper.[20] She was not going to make that mistake.

On another level, however, a crisis at Arizona farm was nearly unavoidable. White farmers represented power in everything they did. Their political participation raised the same suspicion that surrounded their citizenship: rural Euro-Zimbabweans possessed privileges beyond the reach of ordinary, black Zimbabweans. In this case, their property, rather than their heritage, set them apart. Farmers operated through a kind of paternalism that Blair Rutherford (2001) terms "domestic government." On these estates, permanent workers in the compound shopped at the farm store, and often recreated on the farm's football pitch. There was simply nowhere else to go. No public areas separated farms, and little transport existed to bring workers into, say, Macheke town. Whether the farmer wanted to or not, he or she controlled access to and communication with the entire workforce, even on weekends. At election time, the farmer either allowed the MDC to enter or barred it from entry. Harper vowed to "sit on the fence and mind my own business" (Staunton 2005:473), but neutrality was not an option. Moreover, farmers who *did* support the MDC seemed—as the state constantly accused—to exploit unfair, undeserved opportunities. Some donated money, perhaps thereby drawing on fortunes derived initially from the colonial period. More dramatically, one rancher distributed MDC leaflets from his private aircraft (Wolmer 2007:196). Employers turned their class position to political advantage in simply talking to their workers. Prior to the 2000 parliamentary elections, for instance, Johann Swanepoel guardedly presented to his staff "only the facts" regarding trade and foreign exchange. A victory by ZANU-PF, he anticipated, would damage commercial agriculture.[21] Very likely, workers heard in these words a hint of future retrenchments—and weighed their electoral options accordingly.[22] In short, even when they wanted to, Swanepoel and other rural whites could not act politically as ordinary citizens. Through the control of jobs and territory, they exercised disproportionate authority.

In 2000, after a narrow win widely considered unfair, the state resolved to tackle that authority head-on. The party adopted a strategy of co-opting and controlling farmers. In part, this shift grew from the violence paramilitaries were, by that point, wreaking on farm owners nationwide. Armed groups were threatening almost every farmer in one way or another, particularly those identified with the MDC. Some of this violence appeared more public and demonstrative than mere land-grabbing would have required. On August 22, 2000, for instance, the CFU's "farm invasions update" described a particularly theatrical attack in Virginia:

> ... three war vets ... stopped the tractor and told the farm labour to watch while they taught the white man that they were going to get his stubbornness out of him. They pushed him around and frog marched him to show them where the boundaries of the farm are.[23]

Perhaps, in the style of Franz Fanon, such humiliations of the settler inspired support among the natives (Fanon 1963:93). At least, physical manhandling showed everyone that the state was now running the highveld. In 2002, as presidential elections approached, paramilitaries coerced farmers to take part in election rallies. In Virginia, most of those still farming in 2002 joined the party and raised their fists in support at its events—"just for safety purposes," as one farmer explained.[24] Whites occasionally contested their subordination. Shortly before the elections, for instance, ZANU-PF youths demanded that a certain farmer pick up MDC leaflets distributed along the Virginia road. He refused and, according to the CFU's bulletin, "pointed out [that,] if his labourers did pick them up, they would be assaulted and accused of being [supporters of the] MDC." Nonetheless, on that day, workers attended the party's "Star Rally," and fearful farmers provided vehicles to take them there.[25]

In less violent matters as well, farmers simply fell outside the law. The "political" nature of perpetrators and their offenses rendered them immune from prosecution. Theft, for example, ceased entirely to be a criminal matter. Paramilitaries and squatters alike pilfered equipment, standing crops, and personal property belonging to whites. "There is no way you plant among them," said Swanepoel of this new neighbors, "they'll steal you blind. You've got no control."[26] Police, when they came at all, refused to intervene. Even civil servants sometimes treated farm invaders as above the law. In late 2000, an outbreak of bovine anthrax—possibly transmitted through squatters' cattle—brought the animal health inspector to Virginia. He communicated, said the CFU, "that he is unable to intervene as the matter is 'political.' "[27] Such lapses, of course, outraged the farmers, but they soon adjusted their expectations of the state.[28] The withdrawal of insurance—by private companies run or staffed by whites—hurt them more deeply. "Nothing is bloody covered," lamented one evictee, still suffering from injuries at the hands of Stevens's murderers.[29] In fact, insurers were making use of a long-standing loophole in policies. As exclusions from its coverage, the "Farmers Comprehensive Policy" listed losses incurred during "mutiny, riots and strikes, [and] civil commotion . . . amounting to a popular uprising . . . "[30] Popular in some quarters, the farm invasions seemed to qualify. At least, according to the agent who sold this policy to Virginia farmers, "police would just write 'political' on the police reports; so the insurance companies were happy."[31] Officially, there was no longer crime in Virginia, only politics.

This politicization shaded into frank racism but was not reducible to it. "You are another fucking white pig," occupiers shouted at Martin Wiles, a farmer from just outside Harare. His father concludes—in memoirs written after their eviction—"the only motif [*sic*] is racial malice" (Wiles 2005:54, 41). In a more measured voice, James Muzondidya writes of an irrational or self-interested "primordiality" underlying anti-colored, anti-Indian, and anti-white discourses.[32] Indeed, in the first and second instances, the state and war veterans contradicted themselves and could not construct a clear narrative. In the last instance, however, many whites *did* possess historically rooted, if not always primordial, qualities, around which a more careful state could have crafted sensible policies. These characteristics were twofold. First, white farmers owned large holdings, acquired before independence or often with wealth generated before independence.[33] Second, many whites possessed truly essential European birthrights, guaranteeing exit from Zimbabwe and protection elsewhere. This dual economic and national heritage marked white farmers as an exclusive ethnic elite—and a legitimate object of some sort of economic redistribution. Indeed, after 2000, nearly all admitted that land transfers were necessary. It was too late by then. Mugabe himself had labeled white farmers "enemies of Zimbabwe" (Lamb 2006:204; cf. Selby 2006:301). The state's frank racism and unbridled violence rapidly drowned out more temperate discourse on the subject of privilege. Still, if one were to construct such a narrative in retrospect, the invasions addressed entrenched wealth *as well as* a skin color. Although few would admit it publicly, many of my informants understood this dual reason for their victimization—and its irony. Ultimately, the state returned highveld farmers to the class position their ancestors and predecessors had created. The invasions recast Zimbabwean whites as European settlers—minus colonial power!

What Role to Play?

In the face of racism and violence, many whites attempted to reinvent themselves. At a public meeting in Harare, activist Jenni Williams declared, "It's time for us the rewrite that phrase [commercial farmer] as 'people of agriculture.' "[34] On the highveld itself, remaining farmers imagined themselves to be something other than or supplementary to large landowners. The acceptable choices were limited. Farmers did not consider, for instance, taking up smallholdings in communal lands, that is, joining the black peasantry. Nor did they wish to enter the industrial trades of the black, mostly urban, working class. Rather, in the choices they did make, farming men and women recycled three colonial models for white life on the highveld: conservationist, missionary, and native commissioner. Godwin's mother had practiced a fourth acceptable occupation—"bush doctor"—but this profession required more education than farmers possessed. Rather than retraining themselves, farmers built upon their amateur passions. The love of nature, the spreading of the Gospel, and the administration of peasants dovetailed, and individuals frequently pursued two or all three of these options at the same time. Combined or singly, these roles offered whites the promise of respect and symbolic authority on the highveld—while lowering the political profile of the commercial farmer. They also lowered the economic profile by requiring far less capital than dam building. Thrift suited whites well, as they were either losing fortunes or sequestering them overseas. In a fashion barely conscious, then, Virginia whites wished to regain the neutral space they had previously imagined themselves occupying. They hoped to keep their land and keep farming it without *being* farmers. In some cases, national and international organizations helped, but, in the end, such an improbable aspiration could not be fulfilled. The state either evicted these landowners or forced them to reside unproductively on their land. When I met these families during their removal—and afterward in Harare, the UK, and Canada—they seemed to have suffered more for their efforts. They had lost their position on the highveld twice over.

Conservation required the least exertion and garnered the least success. Anglers and dam-builders had already crafted this identity, ironing out many of its inherent contradictions. Could whites now decouple their high-minded love of the savannah from their more mundane possession of it? Steve Pratt, the CFU's representative in Marondera made the most earnest—if not entirely self-aware—attempt of anyone I encountered. As an artist, he disseminated his ideas through lines and shapes, a language that seemed above and outside of politics. In 1998, he published "Kudu Drift," a series of drawings tracing an imaginary farm from the nineteenth to the twenty-first century. He anticipated land reform, and the last two drawings depict alternate futures: "The coming of age" or the "Rape of Eden," as the captions read. The latter image shows severe erosion and crop failure (Figure 5.1). Pratt's caption explains: "Man has forfeited his ability to control the land and his destiny to a perverse and merciless Nature."[35] As late as 2002, when I met Pratt off his farm, our discussion turned to the essential quality of the wild and, as he put it, to "set[ting] aside pieces of land [to] . . . just *be* without human interference."[36] Pratt lasted only a few more months, leaving Zimbabwe for France and eventually England. Three years later, when we shared lunch in Devonshire, he had started painting again. "It's a lot more colorful here actually," he explained, and the countryside's "soul . . . is very deep." Through art, he was adapting. "I can kind of define myself by the landscape really," he explained, adding that he painted topography because "I need that kind of attachment."[37] In the end, love of nature allowed Pratt to become something other than a farmer—but not, at the same time, to continue farming in Zimbabwe.

For those who stayed on or near their farms, the conservationist path caused greater unease. Paramilitaries were destroying flora and fauna with what seemed like joyful abandon. In 2000, war veterans chopped down the gum trees Cathy Buckle had planted ten years

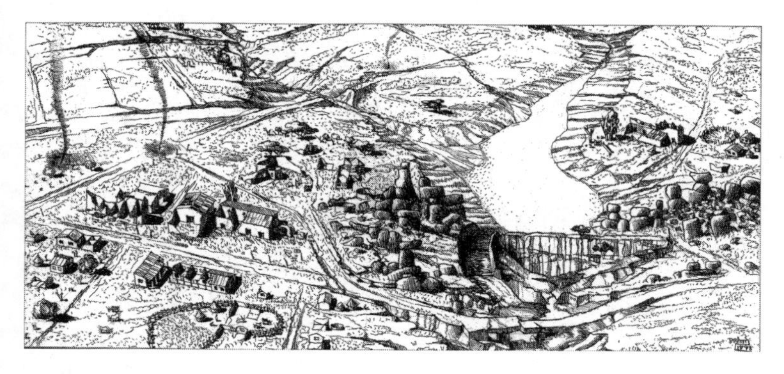

Figure 5.1 "The rape of Eden," drawing by Stephen Pratt, 1998

earlier. Like Pratt, she associated deforestation with sexual violence—in an even more literal sense. "I feel as I imagine it must feel to be raped," she confided to readers of her weekly email bulletin.[38] Then, as I heard in her new home in Marondera in 2002, "they cut the indigenous trees to incense me."[39] Perhaps, such spite—in addition to a need for fuelwood—motivated the occupiers. In any case, conservation-minded farmers detected gratuitous forms of violence against the environment, and they could not tolerate it. Frank Richards, who constructed chalets on his game-rich Virginia farm (see Chapter 4), initially established good relations with the occupiers. Then, the unspoken contract fell apart. "What killed it," he remembered, "was the theft and killing of animals."[40] At roughly the same time, armed poachers shot impala and steenbok in the vicinity. Surprisingly, police intervened, firing ineffectually on the illegal hunters.[41] Meanwhile, in late 2001 and early 2002, occupiers tried twice to ambush Richards. By the following March, the war vet commander on his farm was carrying a rifle openly.[42] Richards and his family soon evacuated to Harare, where I interviewed him. Rather than dwelling on the obvious threat to his person, he reiterated environmental concerns. "I mean," he appealed, "you can't just sit and watch your farm destroyed."[43] In fact, some farmers did just that, but for Richards, Buckle, and other farmer-conservationists, the environmental *was* personal. Even if the paramilitaries had allowed them to stay, they could not have abided their new neighbors. Rather than helping farmers fit into the reconfigured highveld, conservation drove them from it.

Whites' second alternative calling—evangelism—combined human and nonhuman elements with greater promise. As a historical precedent, early missionaries had distinguished themselves from settlers through their good works on behalf of blacks. In 1933, for example, George Wilder retired from nearly 40 years at Mt. Selinda and Chikore Missions confident enough in the people's love to call himself "the white African" (Wilder 1933). More recently, the "lay missionary" Inus Daneel writes of himself, "he has the skin of a *murungu* [white] and the heart of a *mutema* [black]."[44] Could farming whites forge a similarly comforting hybridity? The Stevensons, who had been born again during the stress of the 1970s war, opened a Christian bookstore along the main Harare-Mutare road.[45] Just west of Marondera, they practiced "farming God's way," a set of conservation rules codified by the Harare-based organization Farmers for Jesus. Mr. Stevenson had joined this body, and showed me its literature on "well-watered gardens." Cultivated with zero tillage, such fields combined the green, the pretty, and the rectilinear along principles familiar to the highveld. Plant "where . . . it would be most aesthetically beautiful," Farmers for Jesus recommended to its mostly black adherents. "[T]he Pythagoros theorem . . . is the universal principle of mathematics given by God."[46] Here, a verse on right angles supplemented the early missions' gospel of the plow. As before, agricultural extension would open heathen ears to the Good News. Brian Oldreive, chairman of Farmers for Jesus, was educating peasants and small-scale commercial growers in 15 countries. "I'm now farming a continent," he told me as we walked his tiny test plot outside Harare.[47] In theory, such mass conversion required almost no land base. Practice suggested otherwise. Although Stevenson was, as he put it, "spreading the Gospel to the darker corners of Africa," he still clung to his home farm. "The Lord," he argued to me and to the state in 2002, "forbids that I should give you the inheritance of my fathers."[48] Unmoved by scripture, paramilitaries cleared him out within months.

In Virginia, another member of Farmers for Jesus also failed—but in a different way—to marry evangelism and landownership. Like the guerrilla war, invasions had strengthened the faith of many farmers. By 2003, many of those who remained credited the Almighty. "[O]ur faith in God," averred Mary Fisher, "I reckon that is the pivotal survival point, if you want to write that down." At the request of farmers, her husband, Paul, had read Bible verses daily over the community's radio network. For blacks, the couple had established a Bible school, hiring a pastor and providing construction and materials.[49] Although nominally linked to Oldreive and his organization, the Fishers' evangelism had more to do with giving thanks—that the invasions had not been worse—than with farming four-sided fields. Said Mary, "We feel we have been forced into being missionary by design of what has happened."[50] If, in this fashion, the Fishers counted blessings and saw the cup as half full, they also turned the other cheek. As paramilitaries and other squatters occupied their estate, Silver Cloud farm, the family acceded to their demands. Neighbors disparaged this strategy. "The whole Fisher setup is so loose and so porous,"[51] criticized one. More floridly, recalled another peer, the Fishers "wouldn't say 'fuck off' [to occupiers] when you could say 'fuck off.'"[52] The Fishers had seemingly silenced themselves.

In legal terms, they had secured their position as one of resident managers. In 2003, they actually proposed this status to the

Murehwa District Council. "In response to your letter requesting accommodation . . . as Caretaker," the Council wrote back, " . . . you may occupy the farm house."[53] The rest of the document spelled out caveats and conditions. The Council would withdraw its offer if "the property, the fence, the barns, warehouses, dams, boreholes and other fixed assets are being vandalised." This clause, in other words, obligated the Fishers to protect all of Silver Cloud from misuse by occupiers, paramilitaries, and others. They would have to work for their lodging, and the job would not be easy or overly rewarding. To make sure the Fishers did not settle in, Council reserved the right to revoke the permit, "if Council sees fit to do so." When we met again in 2005, the couple put the brightest face possible on their plight. "We're living in the communal [land]," marveled Mary. "This is a miracle!" Paul just hoped that the situation would not worsen. "If we carry on as we are now," he predicted hopefully, "and we fit into the system, we could stay permanent[ly]."[54] Two years later, the Fishers were still occupying their house, but their economic and social position seemed to have deteriorated. Between trips to establish agricultural ventures elsewhere in Africa, they were trying to gain an interest in a neighbor's intact farming project. This underhanded move appeared to have failed, and friends no longer held them in high esteem. Indeed, Paul and Mary's social status had virtually sunk to that of a *bywoner*, the pitied and despised white tenant in Afrikaner farm novels (cf. Harris 2005:115). In sum, Christian humility allowed the Fishers to inhabit the highveld—but not to cultivate its soil or its society.

The final alternative role for rural whites conceded most to occupiers while, paradoxically, threatening the state more than any other. Some whites sought to provide economic development to interlopers taking their land. Their approach differed from that of postindependence development workers who—in the jargon of 1980 and afterward—answered "felt needs" and fostered "grassroots empowerment." Large-scale tobacco farmers listened less and directed more. They adopted a demeanor reminiscent of the colonial-era native commissioners—among the most syncretic figures that had existed in southern Anglophone southern Africa (Jeater 2007:82). Known as NCs, these officials forged uneasy, paternalistic, often caring partnerships with African chiefs and headmen.[55] In a somewhat similar vein, many Virginia farmers had long provided what they called "over-the-fence support" to the communal lands. They sent tractors to plow, provided tobacco seedlings, and even bought crops from successful small-scale growers. Motivations ranged from the selfless to the admittedly self-interested. "[W]e realized," recalled Debbie Sly, "that we should actually do something about those communal lands... Don't give the fish, teach them how to fish."[56] Fear of theft added to farmers' enthusiasm for development projects. "I'm not going to be able to stay here living in a sea of poverty," worried one farm owner.[57] The precise location of one's farm mattered a great deal. "If you want to farm the communal [land] boundary especially," explained a farmer trying to do exactly that, "you have to cater for your neighbors."[58] By that time, however, smallholders had taken up plots all over Virginia. The invasions brought communal lands to farmers' doorsteps. Indeed, it brought something more desperate than the communal lands. The state provided resettled blacks with almost no assistance, and foreign donors and NGOs refused to be implicated in the land seizures. This institutional reticence created a niche for local whites with expertise in or ambitions for agricultural development. Although none of my informants appreciated this historical continuity, the situation allowed some of them to retool the role of NC.

They did so only implicitly and through Virginia's ultimately futile "community farm plan." Spearheaded by local whites, this vision drew inspiration from a covert, national strategy adopted by the CFU in 2000 or 2001: to consolidate white communities on smaller hectarages and, on the relinquished parcels, to promote agricultural development among resettled blacks. It was a last-ditch attempt to avoid total expropriation, marrying self-preservation with a genuine development agenda. Tom Stone, for instance, set aside 400 ha for himself, retaining control of his house and the dam wall. Other farm owners zoned their properties in a similar fashion, as one put it, "accept[ing] the downsizing so as to secure our infrastructure."[59] Those assets would allow the reduced estates to broadcast development more widely. Virginia's plan and its equivalents devised in other districts promised tillage, inputs, marketing, and other services to resettled blacks. Would this deal give paramilitaries, occupiers, and the state enough of what they wanted? In September 2001, at a summit in Abuja, Nigeria, the CFU presented the plans of every farming district, packaged as the "Zimbabwe Joint Resettlement Initiative." Mugabe's representatives appeared to endorse this proposal and soon passed legislation to enable it.[60] Yet, stronger voices in Harare overruled them. "The powers that be do not want it," Senator Mangwende told one organizer of Virginia's plan.[61] Another promoter, Craig Watson, linked his eviction from Virginia to an overreach on his own part: "We were doing too much. We were doing far too much."[62] By 2007, however, the Watsons had landed on their feet. They moved to Marondera, opened a business in wholesale produce, and established relationships with resettled black horitculturalists. In effect, they were implementing the outgrower schemes of Virginia's defunct community farm plan. Or, as Mrs. Watson confided in connection with the Zimbabwe Joint Resettlement Initiative, "We're doing the same thing, not being resident." "We're there, but we're not there," echoed her husband, trying to sound cheerful.[63] Like the Fishers, the Watsons were living out an alternative model—minus the land that generated the need for such an option in the first place.

In the case of Stanley Hayes, all the weaknesses of the community farm plan and of the unspoken native commissioner model came together. Like the best NCs, he spoke Shona fluently, using it with me—as a means of avoiding a serious interview—for the better part of a year. We finally conversed in English on a sheep farm in Canada, where Hayes had reestablished himself after 2003.[64] Widely respected and admired, he had served on the rural district council with Dave Stevens. Also like Stevens, he had assisted his workers in extraordinary ways, building the first farm school and teachers' housing. Fortunately for him, though, the state initially drew a distinction between Hayes, who had not joined the MDC, and the ill-fated Stevens. "We want you as farmer, Mr. Hayes, we know you're a good man," he recalled the provincial governor promising. Despite such assurances, paramilitaries soon occupied the two farms he owned and gave no indication of leaving. Long before other farmers downsized, he offered 500 acres to the newcomers. "Their response was," he narrated to me, " 'we don't want a little bit. We want the whole lot.' " And Hayes ceded again. In late 2002, he renegotiated with the occupiers for the privilege of living in his house. "My name is worth a lot," he pleaded with apparent effect. "I can do a lot for you if I'm here." Indeed, he plowed for the occupiers, acted as middleman for inputs, and provided transport free of charge. Such services cost only a fraction of the revenue lost through confiscated hectarage. Hayes swallowed his pride. He imagined a form of martyrdom: "[I] have none of my own land and just work for them . . . Hopefully, I'm a Christian man."[65] It was not enough; for Hayes refused to compromise on the one crucial, political point. His eviction followed from a meeting with the Murehwa District administrator (DA) in

2001. As Hayes recorded in private notes, the DA vowed to take his farms:

> "because you are too political . . . Since you became a Councillor . . . you have been asked on several occasions to become a member of ZANU PF and you have stated you never will." I said that was quite correct.[66]

Perhaps, the party would have tolerated a more compliant native commissioner. On the other hand, an outspoken white "whose name was worth a lot" might undermine the party in its emerging constituencies on the highveld. "He could not mince his words," a leader among those occupiers told me much later. "He would not have survived."[67] At last, under intense pressure, Hayes left Virginia and Zimbabwe in early 2003.

In the end, the alternative roles whites implicitly proposed reformulated the old politics of landownership—and facilitated the transfer of working estates. Groping in the dark from 2000 to roughly 2003, farmers experimented with vocations left over from colonial times. Among these, conservation was the least strategic, making the fewest concessions to war veterans. To love wild animals and trees was to wish to protect them—against axe-wielding occupiers. As missionaries, whites adopted an ethic of greater patience and acceptance. Such humility reduced them to tenants on their own land. The final role—that of native commissioner—reflected the deepest deliberation, but it suffered from a central flaw. The state did not want whites to buy blacks' crops and otherwise act as patrons to their poorer, neighbors. The state reserved that role for itself and, perhaps, saw development-minded whites as usurpers. If whites mostly failed to maintain themselves as owners of farms, they at least maintained the farms as long as they could. Not a single landowner followed the Portuguese practice of sabotaging infrastructure, as some whites leaving Mozambique had gone before its independence in 1975 (Hanlon 1984:46). Such a strategy did not even occur to my informants. Hayes rejected any suggestion of it: "I will not destroy anything that we built."[68] He left intact borehole pumps on his property. To have damaged or removed them would have cut against the triple grain of conservation, Christian charity, and economic growth. Guided by those principles, farmers bequeathed as much as they could to their tormenters. *Blinded* by those principles, farmers failed to reconcile their conduct and their values with the constraints of post-2000 highveld politics. As before, they faced and surmounted all challenges except the pivotal, political one.

Playing the Game

By 2003, the balance of forces on the highveld seemed to have shifted. Nationally, most whites had left, emptying whole districts of their farm owners.[69] In Virginia, the initial exodus had slowed, and the families who remained were sorting themselves into two categories. One group was continuing to practice the experiments mentioned above. These farmers would fail and leave the district—or at least cease farming—in the next few years. A second group was adapting successfully. These families would constitute the core community of 11 in 2007.[70] Gaining in stability and profitability, this group developed a strategy of helping its new, black neighbors. These farmers' tactics combined earlier approaches—over-the-fence support and attending party rallies—with a caution bordering on cynicism. Such maneuvers, after all, had failed to secure white land. And whites had experienced a cutting betrayal—that of their workers. At the outset of the invasions, paramilitaries had attacked farm owners and laborers alike, indeed inflicting far more damage on the latter. In defense of their jobs, workers sometimes fought back, ultimately suffering even more assaults. This alliance of convenience persisted until the state implemented Statutory Instrument 6 of 2002.[71] The decree stipulated that, before ceasing operations, farms would have to pay large retrenchment packages to their workforces. This legislation thoroughly fractured the worker-owner alliance.[72] Workers on invaded farms—fearing an imminent closure—demanded immediate payment. Their strikes and other disruptions undermined farmers and, in some cases, may have prematurely terminated viable businesses. Farmers cut their losses financially and emotionally. In Virginia, most shared Peter Farnsworth's assessement of 2002 (or SI 6): "The government's plan to keep the rural folk on side," he explained to me on his farm in 2003, "is to fuck over the white man all the time."[73] When his workers rebelled, Farnsworth paid "packages," laid off the entire staff, and rehired them as independent, unprotected contractors. He described this and other strategies as "playing the game." Landowners proficient in such realpolitik distrusted blacks automatically. They gave only when they had to and guarded advantages whenever possible. Whites then—together with certain blacks—fashioned a new code of conduct on the highveld.

In part, a new wave of black occupants facilitated this reengagement. Having legalized the invasions retroactively—as a "fast track" land reform—the state largely disciplined its paramilitaries (Chaumba et al. 2003). Three new categories of occupants emerged, either replacing or absorbing war veterans. Close to Harare, individuals known as "chefs"—that is, key figures in ZANU-PF, in the armed forces, and in allied businesses—acquired many farms, often for the purpose of weekend retreats. Intent on leisure, they had little need or desire to farm at all.[74] Only one or two such individuals grabbed land in Virginia—on its western edge, closest to Harare.[75] Farther from the capital, this class gave way to more earnest folk. In 2003, the state began to allocate plots to small- and mediumholders who met official criteria. Smallholders, known as A1 farmers, comprised peasants from the communal lands, the urban poor, and, in some areas, former farmworkers.[76] The last category of new occupants, A2 farmers, held medium-sized parcels and derived mostly from the urban professional class. Less politically involved than the chefs—though still known party supporters—they had risen through education and business acumen and hoped now to succeed in commercial agriculture (Selby 2006:328). Their chances were not good, however. The cash-strapped government dispensed little aid, and, because the state provided no tenure security, new farmers could not use their land as collateral for bank loans.[77] In this context of entrepreneurial challenge, then, the A2 farmers—and many of the A1s as well—viewed their white neighbors as an asset, even as potential partners. Whites took this shift in perspective as an opportunity—but not in the generous spirit of the informal native commissioners. Keenly self-interested, the survivors in Virginia apportioned aid in measured, contingent amounts. Farnsworth offered A1 and A2 plot holders what he called "security contracts." They could buy his surplus maize at a guaranteed price, sure to be concessionary in the context of hyperinflation. Although thieves then took half his crop before ripening, the effort seemed to count.[78] In 2007, Nick Mangwende, the head of the association of A2 farmers in Virginia, confided in me: "no one would touch" the Farnsworths. They and like-minded farmers "mix and mingle" by collaborating in agriculture.[79] Such astute engagement won whites allies and, in some cases, friends.

At the same time, the "game" required Virginia farmers to compromise morally, alienating them from many of their earlier friends. Farmers displaced to Harare often treated their peers still cultivating the highveld as collaborators in the worst sense of the term. "[T]hose [of you] who have made deals," someone wrote on a farmers' email list, " . . . should be ashamed of yourselves as you have contributed to the destruction of the Agricultural industry . . . "[80] Bruce Gemmill, Virginia's former ICA chair, spoke with more personal animus. Walter Finch, a well-known Virginian, "has his nose up ZANU-PF's backside," he informed me at his new home in Harare.[81] Farmers in Finch's position accepted criticism but felt they had no alternative course of action. Virginia whites had to trade dignity for security, as became clear at a 2002 farmers' meeting. When the discussion turned to tillage, Hayes laid out his own ethics as a Christian man-cum-native commissioner. He refused to plow for blacks occupying farms adjacent to his own. To do so would have legitimated the eviction of other members of the Virginia community. Yet, on occupied portions of his own estate, he willingly turned the soil for A1 and A2 occupants, indeed charging them market rates for each hectare. Another farmer doubled those fees, but neither tractor owner actually managed to collect payment with any regularity.[82] Even more philanthropically, many farmers grew tobacco seedlings for transplant onto seized farms. A subsequent meeting in 2003 reconsidered this policy. "Make a stand now and say 'no seed beds,' " advised Henry Hart, "what we want to do now is send a message to the hierarchy." In a more cautious tone—and now on the topic of antitheft measures—Roy Baker counseled, "We've got to play the game, but we've got to keep ourselves safe."[83] In such discussions, the specifics of each issue often obscured the larger, psychological point. Compromise was a state of mind. "If your attitude is right," said Swanepoel, "you can get on with these guys."[84] Such familiarity—metaphorically linking white faces and black backsides—only accentuated the social gulf between farmers losing and farmers keeping land.

In 2002, evicted landowners institutionalized this divide by forming Justice for Agriculture. Known as JAG, the organization arose from dissatisfaction with the tactics of the Commercial Farmers' Union. Through proposals such as its Zimbabwe Joint Resettlement Initiative, the Union had struggled to keep farmers on the land. It did very little for them once they left the farming districts. Like Virginia's dwindling white community, the organization encouraged compromise and collaboration—in a fashion that struck many as spineless. In 2002, evicted farmers who had nothing left to lose founded JAG as an explicit, public defense of principle. At a semiofficial inaugural meeting—held at a Harare golf club—the national chairman referred to the state's "assault on title." In even more colorful language, JAG's public relations officer, Jenni Williams, warned against cooperating with land seizures: "appeasement is feeding the crocodile hoping it will eat you last."[85] Within the week, she addressed the media, NGOs, and applied academics, such as myself, at JAG's official inauguration. Speaking at Harare's most elite hotel, Williams denounced the state's "rampant, racial, divisive" legislation. "After the storm," she predicted hopefully, "the sun is definitely going to shine in the Zimbabwean utopia, a free and democratic utopia."[86] More practically, JAG functioned as an interest group of and for commercial farmers. From 2002, it pursued a twin agenda of restoring farms to their owners and seeking compensation for nationalized property. Success eluded JAG on both points, and the leadership settled in for a long battle. In 2005, John Worswick, who had taken over after Williams's departure, described JAG as "mothballing and looking after your title [deed]." The Czech Republic, he reminded me in our interview, was beginning to restore property to owners dispossessed in 1945.[87] JAG would prove whites right even if most were dead by then.

Meanwhile, another group was pursuing an even more confrontational strategy—with equally poor chances of success. International law gave some farmers an extra-African resource. Before the invasions, landowners had registered 70 farms throughout Zimbabwe under Bilateral Investment Promotion and Protection Agreements. Ratified in 1996, this accord facilitated foreign investment by shielding it from any future nationalization. In agriculture, a number of Europeans had availed themselves of this accord to start flower-growing projects in Zimbabwe. Now, these individuals sought to invoke its protection clause. Ben Funnekotter—born in the Netherlands and displaced from Virginia—led the charge. In 2006, his lawyers filed a claim before the International Centre for Settlement of Investment Disputes in Washington, DC. The following year, as the tribunal had finally empanelled judges for the case to be heard in Paris, Funnekotter blustered to the press, "The Zimbabwe government will be responsible for the payment of the claim . . . and they have to pay in the currency of the nationals—which would be euros."[88] Sensible as it was—since the Zimbabwe dollar was inflating at 1,000 percent per annum at that point—Funnekotter's choice of currency also conveyed a symbolic message: it represented an abandonment of Zimbabwe. The litigants were—in Funnekotter's words and in the legal jargon—"nationals" of another country. Back in

Virginia, farmers who could have joined the lawsuit—because of damages to their crops and movable property—avoided it. Some cited the high cost of lawyers. Less explicitly, they might have withdrawn from his effort for political reasons as well. Remaining farmers felt compelled to represent themselves as patriots, devoted to the country and the economy practically and symbolically. Funnekotter's ploy risked portraying them as misplaced, angry Europeans—and it was not paying out euros anytime soon.[89]

If these legal maneuvers achieved nothing material in Virginia, they at least helped farmers to think through their political beliefs. In 2002 and 2003, my informants labeled Justice for Agriculture as dangerously naïve. According to Roy Baker, JAG made the mistake of "think[ing] you can stick your head up vertically in Africa as a white." JAG, he continued, drew its assumptions from a different place. Its notion of law and order constituted a "British attitude."[90] Farnsworth, who was British-born, rejected JAG for the same reasons and faulted it for not appreciating whites' enduring vulnerability. "Good old Africa always kicks you in the balls," he related sarcastically, "so, in ten years time, we're going to have the next Bob [Mugabe] or the next Idi Amin or the next Hastings Banda" who will try to expel whites.[91] With this pessimism, whites slotted themselves into the subaltern position. They shared JAG's faith in white rectitude but retreated from its defense of principle. Instead, they embraced what they saw as tough-minded realism about a flawed continent. "In Africa, you just have to have a humble attitude," said Swanepoel.[92] More pointedly, he lectured me at a farmers' meeting, "You have no rights in Africa. You're a white."[93] If Virginia farmers referred to legal entitlements, they did so only to note their irrelevance. Being alive and on the land was more important than striving for justice. Later, after a wrenching change of leadership in 2005, JAG seemed to appreciate this point of view. Elevated to the group's new board, Bruce Gemmill addressed farmers in a more forgiving tone: "Only those involved can know the line between survival and betrayal . . . [N]early all of us are guilty of compromising our principles . . . "[94] If more farmers had compromised sooner—argued those left in Virginia—a larger number would have continued tilling the land.

Perhaps conciliation came with such difficulty because it cut against farmers' grain in a double sense. First, title deeds contributed to stubbornness, as another member of JAG's board conveys in his memoirs. Squatters have commandeered an orange orchard, and the inspector of police entreats its owner, "Meester Harrisoni, why don't help your new neighbors?" "Inspector, what is mine is mine," retorts Eric Harrison (2006:184). Second, in addition to relinquishing this pride of ownership, those who did help the A1 and A2 farmers had to train themselves for a new form of mental labor: they had to learn to give consideration to blacks. Used to firing orders at their labor force, farm owners now had to palaver. "You never say 'yes,' but you never say 'no,' " explained one farmer. "Always try and have a bit of humor," added his friend in the same conversation.[95] Farnsworth found less to laugh about. After post-traumatic stress disorder and "life-saving" therapy, he considered farming the highveld to be "a mind game . . . a psychological war." "Look," he insisted, "it's a fucking ball-ache but you either accept it and get on with it and play your various games in a day," or you lose your farm.[96] Fortunately for Farnsworth, his "farm" consisted of 3 ha of high-quality flowers grown under plastic. With such intensive production, he could tolerate squatters and other interference on most of his estate. The matter was not so easy for Roy Baker, who grew tobacco on hundreds of hectares. Still, by "go[ing] and giv[ing] a hand," he had won a semi-official reprieve.[97] In 2006, Chief Mangwende vouched for Baker in a letter to government: "He is a man who has proved that he can stay with others well."[98]

Such compromises succeeded, in part, due to shifts in the macroeconomy and related policies. Just as commercial farmers benefitted from the expansion of the 1990s, the shrewdest among them exploited the dramatic contraction after 2000. By 2005, the county's gross national product was shrinking faster than any other in the world and the annual inflation rate had reached four figures. As the Zimbabwe dollar collapsed against all other currencies, the state was taxing and confiscating foreign exchange almost wherever and whenever it could. In this context, farmers needed to earn and sequester U.S. dollars, euros, and so on—and do so with less and less land at their disposal. After 2000, farmers redoubled entrepreneurial efforts begun in the 1990s.[99] They invested tobacco profits—enhanced through irrigation—in the equipment, training, and air freight necessary to grow and ship flowers and high-quality vegetables. Then, through European contacts, these exporters found buyers and hired middlemen. In Holland, for example, farmers' agents personally assembled cut stems for the daily flower auction outside Schiphol, the largest in the world. Underinvoicing and similar accounting tricks kept a portion of sales in Europe. Farmers—who did not discuss family finances with me—started referring to their local accounts as worthless "Shona dollars." Meanwhile, they cut their losses in tobacco, cattle, and all other extensive land uses. Financial cunning permitted territorial generosity. Even landowners still growing conventional crops could give new, black occupants something they needed: access to tractors. The state could hardly provide tillage and was encouraging A2 farmers to make a deal with Chinese interests. A meeting of the Virginia A2 farmers' association—to which Nick Mangwende invited me—considered an improbable offer. As ZANU-PF's provincial political commissar explained, new growers could, by selling their tobacco to a Chinese-Zimbabwean joint venture, earn their own farm machinery. Yet, the politician himself undercut the scheme, referring to its proponents derogatively as "*maChina*."[100] Indeed, the contractual conditions were unfavorable to the point of derision. Whites—known throughout the district—presented a much friendlier face as well as better terms. Even if discouraged by the ruling party, resettled farmers were increasingly prepared to make cross-racial deals. Rapport, then, was building from both sides, based on economic sense as well as on new social sensibilities.

Had Virginia's whites, then, relinquished their environmental escape and become social pluralists? Yes, to a certain extent, their new neighbors disrupted earlier practices of deliberate inattention. In 2005, I greeted Roy Baker disembarking from his fishing boat on the Farnsworths' reservoir. (The cover photo shows this body of water, as well as the Farnsworths' house and boat.) Occupiers had shouted obscenities at him, he related, and because of such "local politicians" he no longer fished from the banks at all.[101] The Farnsworths themselves had never angled with great passion. Yet, after the invasions, the family went more frequently to Kariba. "On a houseboat," advertised Peter's wife, Lisa, "you're free from everything. Nothing will touch you, unless a crocodile. It's not CIO [the secret police]. It's not ZANU-PF."[102] When off their farms, farmers still preferred dangerous critters on four legs to those on two. On their farms, they worried about African bipeds as never before. Even children could no longer imagine the farm as empty savannah. During the same conversation in which she extolled Kariba, Lisa showed me a poem written by her 12-year-old daughter. Entitled "A Farm in Zim," the work begins with a predictable reference to the family reservoir: "My life on a farm is filled with fun/swimming and running around in the sun." Paramilitaries then insinuate themselves—"But every day there is ongoing strife/to keep this lovely sensation of life"—until, close to the end, "It's hard to keep going with all this treason." Here the girl's rhetoric moved from political awareness to criticism, and the Farnsworths felt compelled to edit. They sanitized the text as, "It's hard to keep going *throughout all the seasons*."[103] Like other farmers, the Farnsworths were walking a fine line, placating "local politicians" while hiding from national ones.

All this attention toward blacks did not inspire Virginia's remaining whites to like them any better. Indeed, the shrinking of their community drove whites into tighter and tighter circles of sport and entertainment. In 2002, paramilitaries had forcibly closed the popular Virginia and Macheke Clubs. Farmers then tried to institute regular dinner meetings, some of which I attended until whites themselves discontinued these gatherings. They attracted suspicion from the local government and, in any case, a wave of departures in 2003 pushed the number of attendees below a critical threshold. "Social life sucks," complained a young couple to me in 2005. Many of their friends had fled to town; so they expended precious fuel driving to Harare. Could they not make new, black friends, I asked—at least among the province's smattering of bank managers, businessmen, teachers, and doctors? In an earlier era, explained one of my informants, whites might have made contacts with such people. Few had, however. Now, she continued, "we don't want to step out of that safe zone because we have been out of it for so long."[104] Whites were protecting themselves from those responsible for the invasions. But, I continued in many conversations, most blacks dislike the ruling party, and the educated ones had surely voted for the MDC.[105] In their responses whites' emotions overwhelmed their reason. When I mentioned interracial dating, Farnsworth exploded: "There is no fucking way my daughters are having anything to do with a black man . . . because there is such a lot of pain and suffering in my heart in the last years."[106] In one sense, whites were practicing a form of collective punishment by shunning all blacks (K. Alexander 2005:206). In another sense, they were merely recognizing their limitations. The invasions, many confessed to me, turned whites "more racial" than ever before.

If not always admirable, this particular form of consciousness made certain truths apparent. As early as 2002, Hart criticized his besieged neighbors for failing to appreciate that "they were a minority in a black man's country."[107] Long avoided, the demographic facts loomed increasingly large in white imaginations. This collective self-understanding had two consequences. First, whites took fewer chances in wider society. They knew they could trust whites, or at least trust them not to take others' farms. A given black man, in contrast, might or might not take one's farm. Lauren St. John, who grew up on a farm in Zimbabwe, writes in her memoirs of a similar calculus during the guerrilla war: "there was an unspoken wall between me and the Africans . . . although not all black people were terrorists, all terrorists were black" (St. John 2007:158). On the basis of pigment, whites constructed an enclave of the mind. As the second consequence of minority status, whites developed new practices of venturing outside the enclave. Previously, they had enjoyed "total control," as a soon-to-be-evicted farmer just south of Virginia once described his outlook.[108] Those able to relinquish such mastery—or recognize that it was definitively gone—learned to haggle, coax, wheedle, and sweet-talk. They didn't enjoy such cross-racial negotiations, but many excelled at them so much that they passed the highveld's new political litmus test. Explaining why some whites had been allowed to stay in Virginia, Nick Mangwende avowed, "They did not see color."[109] He could not have been more wrong. Seven years after the death of Dave Stevens, Virginia farmers saw *only* color.

* * *

The practice of *jambanja* recast commercial farmers as colonial settlers. To their credit, a fraction of rural whites then tackled the "settler problem." More frequently, this issue remains unaddressed and unrecognized, as Haydie Gooder and Jane Jacobs (2002:203) argue in their history of white Australians' attitude toward aborigines. The dilemma centers on the way in which former supremacists reinsert themselves into contemporary societies espousing pluralism. In Australia, majority status renders this question one of idle speculation for most whites. Euro-Zimbabweans have never had that luxury. But they acted as if they did. Before 1980, they thought about black society almost entirely in the context of what was known as the "native problem": the administrative project of locating and disciplining black labor. The project of belonging stood apart, inflected toward the landscape. Independence, if it suggested to whites that they try to belong among blacks, did so only briefly and indeterminately. Even the farm invasions initially provoked as much colonial-style paternalism as humility. Prolonged persecution, however, eventually forced whites to see themselves in the cold light of day, as odd men and women out with no historical models. Such a self-assessment did not guarantee continued access to the highveld, but, in some cases, it seemed to dampen the outrage and violence directed against landowners. In Virginia, those who won this reprieve live among blacks, farm among blacks, and engage with blacks in a fashion unprecedented in the history of large-scale white African agriculture. They inhabit a pluralist society—but without enjoyment. The murderers of Dave Stevens and other whites remain at large, liable to take up farms in Virginia or elsewhere. In private, some Virginians fear and loathe, not just those individuals, but blacks in general and sometimes with striking vehemence. Still, such sentiments cause less material harm than one might fear. Euro-African prejudice no longer travels. Whites' political and economic power barely extends beyond the farmhouse door. Moreover, despite racism, farm owners are getting along and forging economic relationships that allow themselves and their new neighbors to survive. Indeed, a certain social guardedness may help them to do so. If, as Paul Carter (1987:163) writes of Australia, boundaries are "place[s] of communicated difference," then whites' boundary building represents a beginning, rather than an end. Hating the farmer next door is one way of acknowledging his presence. A deal across the fence—even an unfair one—locks whites and blacks into coordinated cultivation.

This outcome differs from that in South Africa—both in its level of violence and in the idealism attached to it. Even before the end of apartheid, Rian Malan extols a white couple living in the notoriously destitute and remote KwaZulu homeland. Abandoned by their co-ethnic neighbors in the 1970s, Neil and Creina Alcock became conservationists and native administrators par excellence. They built a dam and invited the men of the district to join hands in a series of agricultural and ecological projects. As Malan describes it, the Alcocks' "willingness to love had carried them deeper into Africa" than any other whites. "It seems that Africa had finally accepted them" (Malan 1991:397, 381). Seeking precisely this sort of acknowledgment for himself, Malan concludes that "you had to live like Africans, until you saw through African eyes, until African problems became your own problems and African pain became your pain" (Ibid.:367). In Neil Alcock's case, the pain became all too real: a Zulu gang assassinated him in 1983. A decade and more later, as South Africa reabsorbed the homelands, elements within KwaZulu-Natal Province still refused to accept whites on the land. Without any state backing or program of land reform, groups and individuals killed scores of white farmers in the Midlands section. "[H]as the battle for the countryside," asks journalist Johnny Steinberg, "been honed down to a lean, zero-sum affair, where every commercial farmer risks his life to keep his farm?" (Steinberg 2002:240). In parts of South Africa, apparently it has. But in Zimbabwe—where racial animosity dates from more recent history and, in any case, fewer guns abound—whites and blacks are finding a positive sum. They share, but they don't quite share problems or pain—or see through the same eyes. They are apportioning the *materials* of agriculture: land, seeds, and farm equipment. Economic survival, not love, motivates this collaboration and perhaps gives it greater staying power.

Will that cold partnership, in fact, survive and thaw into mutual trust? Surprisingly, few observers seem to think that black-white collaboration in Zimbabwe's agriculture can or should continue. In 2005, the University of Zimbabwe hosted a workshop on the land reform program.[110] A panel dominated by agricultural economists considered models for A1 and A2 farming, presumably programs that donors might fund after Mugabe. I suggested that—notwithstanding Mugabe—cultivators were sowing and reaping according to self-help schemes *they* had devised. Indeed, by that point, some whites had begun to manage black-owned farms—in exchange for a fee and/or for the privilege of continuing to occupy their own land. Could these deals represent a model too, I asked? Surely not, the economists responded: the sharing of tractors and seedbeds could not be "sustainable." And, the state had already torpedoed such a proposal after the Abuja summit. Presumably, "chefs" would eventually learn how to run large-scale farms and jettison any unwanted white technicians. Virginians, nonetheless, sustained their relationships until my next visit in 2007. By that time, the currency was nearly worthless, and one-quarter of the country's population had fled elsewhere, chiefly to South Africa. Misery was palpable everywhere. Yet, against the odds, in Virginia, cross-racial agrarian cooperation persisted. It seemed that whites still had a role to play on the highveld. Unheralded and certainly unbidden, landowners and land-grabbers had found a vernacular solution to both the "settler problem" and land reform. And, at last, these white Zimbabweans had escaped from their own trap of environmental escapism.

CHAPTER 6

BELONGING AWKWARDLY

In 1996, Virginia farmer Lee Gavras shot a personal video of his dam under construction.[1] In one scene, Lee's sisters stroll on the nearly complete dam wall. Behind the camera, Lee narrates:

> Check out this dam, *boet* [buddy]—too good. [Intoning] *It's a dam of note!* One day [the sisters] . . . will have their little cottage here, coming on their yearly annual holiday at the dam, the family dam . . . Look at all those birds on the dam wall there, on the side of the bank: those two birds [the sisters] and the birds, the white birds [referring to what appear to be great white egrets, *Egretta alba*] . . .

Reveling in avifauna, shoreline, chauvinism, and sheer pride of ownership, Lee expresses the highveld's Wordsworthian beauty and hydrology of hope at their decadent peak.

In another scene, however, the video breaks entirely with convention. Again at the impoundment and holding the camera, Lee solicits the heterodox views of his Greek-born father, Constantine. "Start. Talk, Speak," Lee requests. The older man turns and, with a determined gesture, recounts for posterity:

> *Constantine*: We built this dam so that we, too, can leave something behind to the blacks. All these years while we have been living in Africa, these people have been taking care of us.
>
> *Lee*: And what do you mean?
>
> *Constantine*: [Taking care of us] so that you can stay in your place [home].[2]

The father's attitude seems to astonish the son. In 1996, after all, no one demanded that whites acknowledge black contributions to their prosperity. Among whites, few anticipated that blacks would take white-owned dams and farms, and fewer still thought blacks might deserve to take them. Even ten years later, Constantine's sense of indebtedness far exceeds the strategic humility proffered by farmers "playing the game." When we met in 2006 in his new home in Texas, Lee explained his father's outlandish sentiments. Constantine, he recalled, had always looked after his workers—through a farm school and similar services. Constantine also knew how they felt. When he emigrated from Greece to Rhodesia in the 1950s, its lily white society treated him as not-quite-European. Ethnic slurs and other forms of discrimination, suggested Lee, dampened any sense of entitlement. Constantine expected little and, in his own mind at least, appropriated little of what came his way. He grew tobacco profitably, of course, but he did not cultivate an ethic of mastery. Long before it became necessary to do so, Constantine belonged *awkwardly* in Zimbabwe.

For Zimbabwe's whites, the imaginative project of belonging has reached a point of rupture, at which it will change course or end completely. In the past, many writers, painters, and photographers felt the need to belong in Africa and creatively imagined a means of doing so. They accompanied and partly steered the white population away from social engagement and toward environmental engagement. Metaphorically and sometimes literally, artists pushed blacks out of sight. The move preserved white identity and, especially among farmers, fostered the fabrication of a secure, geographically grounded, and forceful persona. Lauren St. John writes of her neighbor, Thomas, in the 1970s:

> [E]veryone in the house . . . walked on eggshells around him . . . Even the air seemed to get out of the way for him. For all that, it was impossible not to be drawn to him. His presence was contagious. He had an infectious belly laugh and a deep-rooted certainty of his position in the home and in the country that seemed a throwback to another age—a time when men like Rhodes bestrode Africa, as Thomas still did, asserting their will. (St. John 2007:20)

This sense of mastery—sometimes right down to the body language—withstood minority status and even political ouster. Then, the black-on-white violence of 2000 upended it completely. Sorely outnumbered, whites appreciated the demographic math as never before. Most left in the ensuing years, escaping physically by plane and automobile. For those who have stayed, the fait accompli of black majority can no longer be concealed or ignored. Living with this monolithic fact, then, how will whites reestablish a sense of security and belonging in Zimbabwe? The effort will require renewed ingenuity and liberal borrowing from other Euro-Africans.

Such a deeper reconciliation would reconfigure the bargains underlying Zimbabwe's modus vivendi of 1980. Those implicit deals between whites and the state were twofold. In the first instance, the state maintained wilderness in parks and protected areas as long as it generated tourism and foreign exchange. At Kariba, for example, armed officers kept poor blacks, their crops, and their cattle away from the lakeshore while wealthy whites—domestic and foreign—paid fees, taxes, and wages to use the landscape. This bargain collapsed along with the tourist trade in 2000: arrivals to Matusadona and other national parks plummeted, and the state effectively suspended conservation rules. In the resulting vacuum, the basic structure of a new form of wilderness is already taking shape. Some whites have reimagined nature—in a fashion that will take less space from black production and, therefore, find greater political acceptance. The second post-1980 deal—conjoining land ownership and political silence—will be harder to renegotiate. For 20 years, whites ruled as masters in private space and stood aside as bystanders in public space. That framework exploded when—and partly because—whites entered electoral politics. Now, white farmers master next to nothing. They are living what Alfred López terms "post-mastery whiteness"—and not enjoying it in the least (López 2005:6). "Whiteness just isn't what it used to be," complains one of Melissa Steyn's South African informants (Steyn 2001). Perhaps, if they are to remain, whites in Zimbabwe will come to see that glass as half full. To do so would mean accepting a costly, truly postcolonial deal—relinquishing entitlement in exchange for participation in African social life. This settlement would mark the end of the settler mentality.

NATURE WITHOUT PRIVILEGE

Once robust, the concept of "wilderness" is now withering under empirical and political assaults. The word, as Roderick Nash argues, "acts like an adjective." It is "a state of mind" (Nash 1967:1, 5). Even sober-minded conservationists increasingly acknowledge the chimerical quality of the pristine ideal. As Thomas McShane of the World Wildlife Fund admits, "The Western notion of wilderness does not hold in Africa, because man and animals have evolved together in the continent's diverse ecosystems" (Adams and McShane 1992:xvii). The same is certainly true for women and on five other continents as well. Indeed, across the biosphere, anthropogenic carbon emissions are causing drought, flood, and worse. Rain now falls on the Serengeti, in part, because North Americans drive to work (McKibben 1989). Those concerned to maintain ecological and climatological continuities with the past would do well to persuade the (mostly white) richer countries to burn less coal, oil, and gas. In any case, unadulterated nature no longer obtains on any part of the planet's surface. Politically, as well, the notion of Edenic landscapes holds less currency. It always implied coercion—the use of force by conservationists and their allies to remove from wild spaces the inconvenient people farming, grazing, hunting, and/or living in them. Wilderness had to be *imposed* (Matowanyika 1989; Neumann 1998). And—because of its visual and aesthetic qualities—it had to be imposed over a large footprint. Euro-American tourists intent on viewing wildlife still wish to see *only* wildlife (Dzingirai 2003:256; Suzuki 2007:231). In catering to this intolerant "arrogance of anti-humanism," managers of national parks have swept local human populations from the visitor's view (Guha 1997). Mostly white elites displace and replace poor, brown, and black peasants. In the end, it is surprising, not that traditional parks are losing legitimacy, but that they still retain any at all.

Much of that staying power surely derives from the more symbolic aspects of white privilege. In the neo-Europes, parks and other conservation areas frequently signal the era of European exploration and conquest. They operate in code. A 1963 report of the U.S. Park Service, for instance, directed managers to "restore the land's original conditions." The phrase might have suggested a Yellowstone or Yosemite congealing as the earth cooled. To its (mostly white) readers, however, the text clearly referred to terrain as witnessed by the so-called explorers: such men advanced westward into a pristine hinterland untrod by Indians or anyone else.[3] North American ecotourism, then, frequently enacts a mythology of white exploration. In southern Africa, ecotourism also memorializes whites' history. The region's flagship park bears the name of the Transvaal Republic's hero-president, Paul Kruger. In the 1920s, as Jane Carruthers writes, conservationists rallied support for the new protected area by "stressing the common heritage and values which wildlife represented for whites . . . " (Carruthers 1995:62). Much later, in 2000, the governments of South Africa, Zimbabwe, and Mozambique vowed to extend Kruger north and east to form the Great Limpopo Conservation Area. Promoters, including Peter Godwin, envision an even more expansive "ecological Cape to Cairo dream" (Godwin 2001:17). The reference to Cecil Rhodes and Britain's colonial ambitions is not accidental. Contemporary conservation dabbles in nostalgia for the colonial period (Hughes 2005:174). Outfitters in Zambia will fly clients in a Tiger Moth, the pre-WWII aircraft in which Karen Blixen fell in love with Kenya. One can stay in the mansion of a colonial knight or dine on the Zambezi attended by smiling waiters.[4] If the pioneer moment lives through North American outdoor recreation, Empire still twinkles in Kipling's "land washed with sun." And promoters wonder why national parks attract so few of even the most wealthy nonwhites: popular conservation continues to produce the aesthetics, symbols, and fables of white privilege.

One may undermine privilege in two ways: by expanding it or by contracting it. Both strategies would push conservation toward greater justice and equality. The expansion of privilege confers its benefits on a larger and larger population—to the point where those advantages lose their exclusivity. Kariba shows a possible, logical way forward. Conservationists initially treated the dam wall as an intrusion, a crime against nature. Recall Reay Smithers's condemnation of it as "the greatest environmental upset ever to befall a population of animals and birds within the African continent . . . " (Chapter 2). His criticism placed engineers, builders, and their creations beyond the pale of conservation. A later generation of nature lovers, however, accepted the artifice of hydropower. Indeed, writers and photographers of the 1980s and afterward have consistently celebrated the reservoir as a waterscape ideal for recreation and visual appreciation. They enlarged the circle of privilege, admitting a culture of hydro-industry into the fold of "nature." Having done so once—and without due reflection—conservationists could now deliberately open their gates again to a much larger class of people. (Black) smallholders, herders, and fishers, after all, damage the Zambezi Valley far less than the river's impoundment ever did. Conceivably, an agro-ecology of cattle and grain can develop in ways that are at least as valuable as a hydro-ecology of tiger fish and buffalo. Elsewhere in southern Africa, this spirit of open-minded experimentation has already inflected conservation. In Zimbabwe, UNESCO has recognized the Matopos National Park as a World Heritage Site—based largely on Ndebele religious beliefs and practices there (Ranger 1999:288–289). In South Africa, private authorities have informally acknowledged the Transkei seashore as a cultural heritage site. Discarding the early label of "wild coast," tourist firms now associate the destination with Nongqawuse's 1856 prophecy that blacks would drive whites into the sea.[5] Times are changing! The "anti-racist environmentalism . . . [transcending] instinctive romanticism," that Bruce Braun advocates for Canada is actually emerging in southern Africa (Braun 2002:212). This more democratic approach values all contributions to the "second nature" of Africa's lived-in landscapes. In exploding the wilderness idea, such hybrids of nature and society extend whites' unearned, favorable dispensation to blacks.[6]

A second, diametrically opposite means of undermining privilege shrinks it to the point of irrelevance. A postage stamp – size park would hardly interfere with peasant production. But would it do the job of a park at all? Putting aside the question of scale and biodiversity, Lilliputian conservation upends the conventions of African tourism.[7] Micro-parks lack long-range, photogenic prospects. Only a less visually obsessed visitor—one graduating from European Romantic sensibilities—would appreciate this nature-at-hand.[8] Again, at the edge of Zimbabwe, experiments have been taking place. Downstream from Victoria Falls, 23 rapids thrash paying customers, flipping many into the river itself. They experience wilderness and savagery up to and including lethal force—all on a watercourse less than 50 m wide. And this adventure takes place on the outskirts of Zambia's fourth-largest and fastest-growing city, Livingstone. The metropolis, though, lies out of sight of rafters, blocked by the lip of the Batoka Gorge. Indeed, the gorge enables a whole new geometry of landscape—of the vertical rather than the horizontal. Operators in the new field of "adrenaline tourism" send their (still almost exclusively white) clients down the gorge attached to bungee cords or rappelling ropes. They experience the landscape viscerally and tactilely—or, actually, not tactilely at all for those who free-fall. At Batoka Gorge, one Zambian (white) photographer has circled back to the visual. Steven Robinson pioneered the use of the panoramic lens in the region's landscape photography, capturing, as he puts it, the "w_i_d_e view of Zambia's stunning wild landscape." At Victoria Falls, he turned his camera on its side. Shot in 2004, Robinson's Batoka collection features vertical strips of sky, rock face, and churning water, conveying the raw energy of a locale Robinson admits "is hardly remote or unknown" (Figure 6.1).[9] But micro-nature is not democratic: only elites can afford rafts, ropes, and framed prints. Still coded white, this concentrated pocket nature at least displaces no one.[10]

Taken together, the mestizo of nature-society and the diminutive of micro-nature represent an unheralded compromise—and hint at more still. The optics of popular conservation are changing. Photographers, operators, and tourists will now settle for a short-distance gaze or tolerate human elements in their long-distance gaze. Surprisingly few public spokespeople in conservation—let alone in tourism—recognize this shift away from Edenic landscapes.[11] If taken seriously, though, these concessions could transport conservation a long way. Small-scale agriculture unites the hybrid and petite ecologies. Could conservationists come to value—not just tolerate—cultivators tilling biodiverse soils? Consider the sentiment of Pearl S. Buck's novel *The Good Earth* (1931). "There was only this perfect sympathy of movement of turning this earth of theirs over and over to the sun," she writes of a Chinese peasant couple,

> this earth which formed their home and fed their bodies and made their gods. The earth lay rich and dark, and fell apart lightly under the points of their hoes . . . Each had his turn at this earth. They worked on, . . . producing the fruit of this earth—speechless in their movement together. (Buck 1931:33–34)

Among the white Zimbabwean texts considered above, there is no equivalent passage. Perhaps this tactile, toiling oneness with the land lies outside contemporary whites' experiences. So many, after all, made a fetish of the bird's-eye, rather than the worm's-eye, view. Perspectives shift, however. In Zimbabwe and beyond, conservationists may soon come to appreciate people who mix, in every way, with the landscape. This postwilderness structure of feeling would, at last, put a human face on conservation.

Figure 6.1 "Victoria Falls from below," photograph © Stephen Robinson
Source: www.spirit-of-the-land.com, 2004.

WHITENESS WITHOUT MASTERY

To anticipate another sea change, Euro-Zimbabweans stand on the cusp of post-mastery whiteness. An ethos of mastery, argues Richard Dyer, inheres in white people generally. Whites, he suggests, explicitly value "the control of self and the control of others," including of territory (Dyer 1997:31). To racial Others, they seem to terrorize and destroy (hooks 2009:95). In the United States—whence both these characterizations spring—some whites exercised mastery continuously, devastatingly, and on a grand scale. From (so-called) discovery, through conquest, genocide, enslavement, and westward expansion, white societies have dominated the midsection of North America (Roediger 2002:132). Of course, few living Euro-Americans have participated in any of these activities. Only a minority even descends from early settlers. But, many still identify with deeds—now national myths—done in their name. White Americans, Dyer argues, embrace a political culture of dominating spatial and social Others. So have the lightest-skinned in South Africa, where, according to Zine Magubane, "whiteness is defined above all by the superior economic and political power that it commands . . . " (Magubane 2004:144). Even liberals cannot escape history. Zimbabwe's past, though, contains as much illusory as material white mastery. The European-derived minority, in fact, controlled very little, even when it ran the state. Many aspects of both blacks and the landscape lay outside its grasp. Perhaps for this reason, Euro-Zimbabweans' imaginative project ignored the former and obsessed about the latter. Any new form of whiteness, transcending mastery, would break both these patterns, directing the imagination away from nature and toward society.

In Zimbabwe, some commercial farmers have already shifted their gaze and given voice to a hesitant humility. In 2000, they lost control of property, production, and their very persons on the highveld. Most departed for town, but a small rump has been able to chart a course through this uncertain world. In part, they have done so by intensifying production, farming their own micro-spaces of flowers and horticulture. The greenhouse is to the tobacco field as the river gorge is to the national park. Some commercial farmers have also intensified their social practices. They have done so by fishing less and palavering more with their unwelcome neighbors. They aspire, not to implement top-down control, but to maintain a balance of forces. Literature, too, has already begun to grapple with farmers' fall from mastery. Bookey Peek's memoir—*All the Way Home: Stories from an African Wildlife Sanctuary* (2007)—begins with the conventional love of nature suggested in its title. In 1989, the Peeks bought a ranch in the Matopos Hills, outside Bulawayo, and soon adopted all manner of injured animals and birds. "[S]uddenly, it was home as if we always had been there," the author recalls. "In the years that followed, we nurtured, protected and filled our little patch of Africa with all the creatures to whom it rightfully belonged." African people apparently did not qualify. "But now," she continues, "...our land was under threat forcing us to acknowledge what we had always known in our hearts, that neither fence, nor sanctuary, nor all our wishing could keep the world at bay" (Peek 2007:323). Peek's realization falls short of the humility of feeling that she never should have owned her estate—of sensing that she never belonged. It lacks the rightness conveyed by Gavras in the family video: that blacks *deserved* the land.

In print, this exceptional sensibility rises closest to the surface in Ian Holding's extraordinary novel. As mentioned in Chapter 1 of this book, *Unfeeling* centers on a boy, Davey, trying to avenge the murder of his parents by war veterans. Narrated in flashbacks, the book only ties its loose ends together at the end, where it juxtaposes Davey's past and present attitudes. Alongside the aptly named Broadlands reservoir, he remembers his childhood:

> he's sitting here at dawn... a gradual brightness rising over the mirrored dam, the wide woven blanket of tobacco fields, revealing the landscape, the savannah, Africa... He has his feet planted in the soil, and even as young as he was then, he knew, one day, he would be master of all that, of all the glory that surrounded him. (Holding 2005:231)

But, things have gone awfully wrong—worse than in the actual invasions, at least as far as they are publicly known. A paramilitary gang has murdered Davey's parents and moved into his house. Davey has vowed to kill these assassins. In reality, rural whites exercised remarkable self-control in an effort to avoid provoking the state into using overwhelming force. Gun-owning white men mastered their own emotions. The young Davey, on the other hand, acts upon his grief and anger—giving shape to white fantasies of revenge. Yet, he and the neighboring farmers, who come later to finish the job, fail in the most emasculating fashion. The black cook poisons the family's murderers even as Davey stalks them. The boy fires his rifle into their corpses. Unaware of this anticlimax, the white men of the district approach, heavily armed, only to be stopped and diverted by a request from the farm labor: they need gasoline to burn the bodies and dispose of the evidence (Ibid.:220–227). Accustomed to acts of heroism, these estate owners serve only as sidekicks. Davey closes the novel "lonely and regretful and unappeased . . . [H]e's just a tiny speck of nothingness lost against the vast, unfeeling wilderness" (Ibid.:243).

Compared to this isolation, recent South African literature sketches white futures both more pessimistic and more optimistic. Writers describe a country at the same time more violent and more hopeful than Zimbabwe. At the most dystopic, J. M. Coetzee pushes to an extreme his observation cited earlier in this book: "that the ultimate fate of whites . . . depend[s] a great deal more urgently on an accommodation with black South Africans than on an accommodation with the South African landscape" (Coetzee 1988:8). Seemingly ignorant of this fact, Lucy—the leading white female character in *Disgrace* (1999)—moves from Cape Town to the interior. She takes up a smallholding and cultivates her garden, surrounded by black neighbors. Then—in a passage that aroused controversy across South Africa—some of them rape her. Afterward, Lucy tells her dumbfounded father that she will remain on the land come what may: "what if that is the price one has to pay for staying on? Perhaps that is how they look at it; perhaps that is how I should look at it too . . . Why should I be allowed to live here without paying?" But she doesn't intend to suffer rape again. The real price she will pay—and pay every day—is a formal marriage to her closest neighbor and the uncle of one of the rapists. He will then hold the land. "Slavery," denounces her father. "They want you for their slave." "Not slavery. Subjection. Subjugation," answers Lucy.[12] This is one form of social integration, wherein whites slot into the bottom strata of African kin and class hierarchies. Antjie Krog undergoes the same sea change as Lucy—but with a more positive outcome. *Country of My Skull* (1998), her memoirs as a journalist (and an Afrikaner) covering the Truth and Reconciliation Commission, frequently falls back on an unreconstructed environmentalist identity. "This is my landscape," she writes of the Free State. "The marrow of my bones. The plains. The sweeping veld. The honey-blond sandstone stone. This I love. This is what I am made of" (Krog 1998:277). Without disavowing that statement, Krog tempers it with a sense of her unease in liberated South Africa. When overwhelmed by accounts of torture, she writes, "I belong to that blinding African heart. My throat bloats up in tears . . . for one, brief, shimmering moment, this country, this country, is also truly mine" (Ibid.:338; cf. Horrell 2004:775). Like the fictional Lucy, the actual Krog finds a route to belonging *socially* in integrated Africa. Blacks will not subject her exactly, but—through their suffering and her tribe's responsibility for it—they reduce her to the role of a supplicant.

These relationships fall far short of harmony or mutuality. In entering into them, white characters relinquish privilege and mastery. Such acts do not abolish prejudice and hate—in whites or in anyone else. Indeed, if the real-life farmers of Virginia, Zimbabwe serve as a guide, the rough micropolitics of face-to-face negotiation only stimulates dislike. As compared with the period before 2000, highveld whites bargain with more blacks and befriend fewer. Does this shift represent progress? The answer to that question hinges on the distinction between humiliation and humility. Suffering the former—as Zimbabwean whites certainly have—does not necessarily lead to the latter. Fanon hoped that violence and degradation would drive the colonizer out, as occurred in Algeria. What if colonizers stay? They can do so only—or most sustainably—if they take steps toward humility, if they accord recognition and respect to their black interlocutors. Recognition lies at the heart of what Kwame Anthony Appiah terms "cosmopolitanism," an ethic in which "we have obligations to others in the broadest sense." Those obligations do not require us to enjoy the company of—or associate at all—with people who are different (Appiah 2006:xv, xx). One might satisfy the obligations strategically rather than generously. "Cosmopolitanism," as Appiah continues pragmatically, "is the name not of the solution but of the challenge" (Ibid.:xv). In Zimbabwe, whites have confronted a particularly dangerous form of that challenge. Now, in the shadow of state terror, some are crafting a remedy to the structural violence of their own past power and enduring wealth. They are establishing what Alfred López calls "a post-colonial *Mitsein*, this being-with others after the fact of domination" (López 2005:6). That is enough—and difficult enough—for now. African whites are, at last, figuring out how to exist alongside and in engagement with the black society around them.

* * *

This book has treated Euro-Zimbabweans as objects of both criticism and charity. For their privilege and prejudice, white Zimbabweans have received ample criticism. When that privilege has evaporated, the prejudice threatens others less. Power fades to delusion. Beginning in the 1970s, the Rhodesian Front underestimated the guerrilla movements—just as so many commercial farmers subsequently ignored black politics. But, such failures in judgment do not render their agents irrelevant. From the wreckage of white institutions, I have tried to salvage a useful picture of mentalities not so foreign after all. Other commentators—including white Zimbabweans themselves—may take this charitable attitude a step further. My informants frequently described themselves as "the Jews of Africa." Referring to prewar Eastern Europe, they imagined themselves as an ethnic minority—unassimilated and, therefore, perpetually vulnerable. More tangibly, a shocked Peter Godwin learns midway through *When a Crocodile Eats the Sun* that his father emigrated from Poland to Rhodesia *as a Jew*. He free-associates to the greater number who settled in Israel. "[T]he muscular sabras trying to reestablish a home in an unforgiving land surrounded by hostile Arabs," he writes, "[their story] resonated too closely with my white African narrative" (Godwin 2007:128; cf. J. Taylor 2002:23). In other ways, the two histories diverge: most obviously, white Africans endured no Holocaust. Still, they have suffered enough to warrant a more sympathetic interpretation than in the past. Some deserve frank admiration for the way in which they have surmounted the explorers' heritage of control. White Zimbabweans are now exploring a possible future for neo-Europeans everywhere.

If whites are thus graduating to post-mastery, they are also heading toward what one might call post-belonging. To make another unlikely comparison, Africans transported across the Atlantic in chains lost both a homeland and the freedom to make another homeland. This terrible uprooting confers one advantage. "[C]onsider what might be gained," encourages Paul Gilroy, "if the powerful claims of soil, roots, and territory could be set aside." One might create what Gilroy calls "planetary humanism" and "placeless imaginings of identity" (Gilroy 2000:111). As their project of imagination, abolitionists, Pan-Africanists, and black writers created far-flung, deterritorialized loyalties and networks. Extra-European whites, of course, have followed a dramatically different path to reach this point. Whereas Africans entered new continents against their will and as property, whites willed themselves to be the property owners of new continents. Africans served as commodities of global capitalism; Euro-Africans increasingly drive global capitalism by producing and selling commodities (Schroeder 2008). Still, in the currently expansive use of the term, both groups live in *diaspora*—in the Black Atlantic or in what one might call the "white tropics." Movement, not fixedness, makes this form of consciousness. A topographical identity fits poorly. "I'll never have the sense of security that this land sees and recognizes me as her own," confesses a nonaboriginal Australian to the landscape writer Peter Read (2000:222). Such modesty fosters understanding, tolerance, and pluralism. In the region of this book, Africans have long welcomed strangers, and the latter have long adapted to their hosts. Oral history of the Zambezi basin indicates that in-migrants frequently adopted the spirits and sacred areas of long-standing residents (Kopytoff 1987). Precolonial travel of this sort bred humility rather than hubris. Perhaps, in today's plural, postcolonial societies, the hesitant newcomer, rather than the confident pioneer or founder, again provides the best role model.

Sadly, the people described in this book will find such speculation idle, if not cruelly insensitive. State-sponsored violence and economic collapse dominate political life and thought in Zimbabwe.[13] Most whites have emigrated, mainly to the sister neo-Europes of Canada, Australia, and New Zealand. The rump in Virginia, Kariba, Harare, and other locales may soon join them. Still, the wisdom obtainable from white Zimbabwean experiences does not depend upon their continuity in situ. Logics of white privilege and environmental escape flourish elsewhere. In Australia or North America, extreme geographical sensitivity—for outback, prairie, and savannah—facilitates an insensitivity to native people. Autochthons of the United States have slipped into a particularly deep obscurity. American "nativist" movements protect the spaces of *whites* against more recent immigrants. Britons, in other words, romanced the land and naturalized themselves in new worlds. But, in so doing, they hobbled imperceptibly their ability to adapt socially. "Wordsworth [is our] . . . burden of the past." Or so Coetzee implies when he slips that phrase, as a book title, into *Disgrace* (1999:4; cf. Barnard 2007:37). The wisdom of white Zimbabwe argues for, at least, feeling the load of such burdens. Freighted in this way, a more provisional form of belonging would acknowledge whites' shallow history and consequently limited knowledge in much of the extra-European world. Hesitation and contingency—rather than fierce certainties—would also serve Euro-Americans well as immigration from other continents pushes them toward minority status. Belong awkwardly, white Zimbabweans may one day counsel their American peers.

Notes

Chapter 1

1. Said (1993:14). Regarding "structures of feeling," see Thompson (1963:194) and Williams (1977:132–135).
2. Lessing (1958a:700). Hughes (2005:160–161) discusses the use of a continental geographical scale among white writers and conservationists.
3. Analyses of colonial administration are too many to enumerate here, but any list should include the divergent accounts of Mamdani (1996), Mitchell (1991), and Myers (2003).
4. Such bifurcations—often less pronounced—obtain elsewhere in the region. Ferguson (2006:121), for instance, distinguishes between Zambia's macroeconomy and the *Chrysalis* periodical, whose intention was "to promote and build a new Zambian identity."
5. Until 1964, the colony was known as Southern Rhodesia.
6. White populations stayed below 1 percent in Kenya and 20 percent in twentieth-century South Africa (Crapanzano 1986:xiv–xv; Godwin and Hancock 1993:287; Kennedy 1987:1).
7. Burns (2002:5), McCulloch (2000), and Pape (1990) address these irrational outbreaks. If a white woman and black man were involved, allegations ranging from burglary to "insults" could easily generate a charge of attempted rape (Kennedy 1987:141–146).
8. Here I differ with Ranka Primorac (2006:68), who describes the "Rhodesian chronotope" as a set of assumptions treating rural space as black and urban space as white. In fact, much of this literature appropriated all spaces for whites.
9. Stockley (1911:167; cited in Chennells 1982:144). Chennells (1982:143) distinguishes Stockley as the first novelist to have spent a significant portion of her adult life in Rhodesia. In the novel, Leander Starr Jameson, who organized the settlement of Rhodesia, makes the statement quoted.
10. See McClintock (1995:24ff) for further discussion of the "porno-tropics."

11. The violence, danger, and masculinity of hunting appear to create similar opportunities (e.g., Ruark 1962).
12. See, for instance, Nancy Jacobs's (2006) discussion of the ways in which colonial ornithologists underrepresented their African field assistants in print. Even those who worked intimately with blacks offered few opinions on them—positive or negative—in the public narrative ensuing from such work.
13. Huxley lived at Thika, east of Nairobi and just outside the more hospitable white highlands.
14. At their height, glaciers reached to roughly the 50th parallel in Europe and the 40th in North America.
15. See Blackbourn's (2006) recent environmental history of Central European postglacial wetlands.
16. Williams (1973:127). For wider-ranging treatments of the importance of water in Western thought and symbolism, see Orlove (2002:xi–xxvii), Raffles (2002:180–182), and Schama (1995:245ff).
17. See Elkins (2005:10, 378 fn. 20) for the derivation of the first phrase. Kipling (1928:239) used the second phrase in his elegy to Cecil Rhodes, "The burial," read at Rhodes's funeral in the Matopos in 1902 (Ranger 1996:169 n. 1). Rhodesia's National Federation of Women's Institutes (1967) recycled Kipling's line as the title for a publication celebrating Rhodesia in 1967 (Chennells 1982:160).
18. See Schutz (1972) for an application of Hartz's (1964) fragment thesis to Rhodesia. Chennells (1982) argues that Rhodesians frequently identified themselves as preserving true British character, even as it suffered corruption and decline in the homeland.
19. Chennells (1982:229). For a more detailed classification and mapping of "grassland," "shrub savannah," "tree savannah," and "savannah" woodland, see Wild and Fernandes (1968).
20. Lessing (1951:49–50). For fuller, richer explanations of this short story and related themes, see Hotchkiss (1998) and Wagner (1994:181–183). "Veld" derives from the Afrikaans for "pasture," and is used in English to mean "bush."
21. Here, I disagree with Jeremy Foster (2008:3), who describes the "subcontinent's spectacularly scenic environment" as facilitating—rather than retarding—white identification with it.
22. Coetzee (1988:62). In the same self-questioning vein, Afrikaans author and anti-apartheid critic Breyten Breytenbach (1996:108) continues the quotation at the outset of this chapter with " . . . We had to go on writing ourselves out there to fit a tongue to the mouth.

And then we lost the language. Are the lines not also nooses?" See also Ashcroft, Griffiths, and Tiffen (2002:135).

23. Lamb (2006:137–138). Lamb's italicized text indicates a quotation from Nigel himself.

CHAPTER 2

1. Lessing (1958b:199). Regarding the dam itself, Lessing (1956) condemned unequivocally both the exploitation of African laborers in the construction and the forced relocation upstream.
2. Ranked behind the Nile, Congo, and Niger rivers, the Zambezi is 2,660 km long and drains 1,330,000 square km.
3. Measured in surface area, Kariba was the largest reservoir until Egypt's Aswan High Dam created a larger one. In capacity, Kariba has always been the third-largest reservoir in the world.
4. Now the Zambezi River Authority.
5. Advertisement for Marion Power Shovel Company, reprinted in Gillies (1999:62).
6. The quotation derives from a tourist brochure of the late 1990s (Murphy n.d.:1).
7. Livingstone (1865:324–325). According to many accounts, Kariba is a corruption of "Kariwa." McGregor (2000) and J. Moore (1965) comment on Livingstone at Kariba.
8. Dempster did apparently exist and was remembered by Ian Nyschens, who hunted in the Zambezi Valley beginning in the 1940s (Nyschens 1997). Interview, Harare, July 16, 2003.
9. In 1959, a South African publisher compiled a book on Kariba (Anonymous 1959), but this volume sold few copies. Gillies (1999) reprints that volume as an addendum to his own text.
10. Clements (1959:13). Jarosz (1992:110–111) presents a similar interpretation of Clements and related authors.
11. Balneaves (1963:opposite 64). Howarth, who read Colson's and Scudder's manuscripts, still describes the Gwembe Valley as "perfectly primeval" (Howarth 1961:vi–vii, 21).
12. Regarding restrictions on agriculture, see Malasha (2002:178–179) and Bourdillon, Cheater, and Murphree (1985:15–25). References on contemporary economic conditions among valley Tonga resettled in Zimbabwe include Reynolds (1991:19–20, 27–31), Dzingirai (2003:248–249), and World Commission on Dams (2000:37ff).
13. The figure usually given, 5,000 animals, excludes those rescued on the Northern Rhodesian side of the lake (Kenmuir 1978:25).

14. See Robins and Legge (1959:opposite 48) and the covers of Lagus (1960) and of a later biography of Fothergill (Meadows 1981).
15. During the rescue, game officers also developed techniques of tranquilizing—notably with the drug M99—and translocating large mammals. In the 1990s, those same methods enabled the stocking of private conservancies and of depleted protected areas in Zimbabwe, South Africa, and elsewhere.
16. Balneaves (1963:159) writes, "not all of man's vast and complex schemes for his own advancement can cancel out the trail of suffering left behind."
17. Genesis 9:13–15; quoted in Robins and Legge (1959:175).
18. Some academics embraced these utopian dreams as well (see Cole 1960, 1962; Reeve 1960).
19. In this sense, the myth is a falsehood that plays the same rhetorical role as the true story of a glacial lake on the Columbia River, roughly where the Grand Coulee Dam now sits (R.White 1995:57). The true geological history of the mid-Zambezi valley goes as follows: The proto-upper Zambezi flowed into the Limpopo valley until roughly 5 million years ago. Uplifting trapped the water in what is now northern Botswana, where it did form an enormous lake. Between 3 and 5 million years ago, the lake overflowed into the proto-lower Zambezi valley (Main 1990:5–8).
20. Interview, Bristol, UK, February 24, 2003.
21. *Hold My Hand* became an international best-seller. Davis published a sequel in 1984 and 11 additional paperbacks. In 1972, he released *Operation Rhino*, his only work of nonfiction, concerning the translocation of an endangered population of black rhinoceros. Although this book made him Rhodesia's first popular conservation writer, he is best known for *Hold My Hand*.
22. This concept—an alternative to African nationalism—underlay the 1953–63 Federation of the Rhodesias and Nyasaland.
23. Interview, Coin, Spain, January 13, 2004.
24. For a similar interpretation, see Chennells (1995:111).
25. Balon and Coche's (1974) less romantically entitled *Lake Kariba: A Man-Made Tropical Ecosystem in Central Africa* sold a good deal fewer copies, as has Moreau's (1997) ecological volume.
26. Interview, by telephone, June 7, 2004.
27. Interview, Fishhoek, South Africa, May 17, 2004.
28. Interview, Muizenberg, South Africa, May 20, 2004.
29. Wannenburgh spent only four days at Lake Kariba, most of it in the company of the noted tour operator and photographer Jeff

Stutchbury (cf. Stutchbury 1992). For recent research on lakeshore *Panicum*, see Skarpe (1997).

30. Wannenburgh (1978:22). Writing in a 1991 photo book, Mike Coppinger and Jumbo Williams echo that sentiment: "We regard Kariba as a successful enterprise . . . the environment apparently has not suffered. The area adapted and retained most of its wildlife and some species [especially *Panicum*] have positively blossomed" (Coppinger and Williams 1991:107–108). Yet, they later contradict themselves: the Botaka Gorge dam planned upstream of Kariba "will mean the irrevocable loss of another wild sliver of Africa, the further disfigurement of a magnificent river. Man's concrete masterpieces will never match the splendours of nature" (Ibid:113).
31. Interview, Muizenberg, South Africa, May 20, 2004.
32. Wannenburgh (1978:27). See Teede and Teede (1990:50) for an almost identical narrative.
33. *Tantalika* was republished in 1984 by MacDonald Purnell (Johannesburg) and in 1990 and 1999 by Baobab Books (Harare).
34. Rayner (1980:34). "Fura-Uswa" probably derives from the Shona legend of Guruuswa, a place of "long grass" north of Zimbabwe from where the people migrated in the Iron Age (Beach 1980:62–63). This tremendously oblique reference constitutes the only engagement between Kariba literature and African understandings of landscape.
35. Interview, Harare, June 30, 2003.
36. At its height, in the early 1990s, the Zambezi Society had roughly 1,000 members, including roughly 100 blacks (Interview, Harare, May 20, 2003).
37. In the magazine of the Wildlife Society—another nearly all-white conservation group—Pitman (1983b:10) excoriated the "insidious attitude that 'Mana has already been killed by Kariba, so why bother with it any more?.' "
38. For a critical assessment of the idea of wilderness in American thought, see Cronon (1995).
39. Interview, Kariba, July 2, 2003.
40. Margaret Peach, personal communication, June 19, 2008.
41. Timberlake (1998:60–61) summarizes the scientific literature regarding alterations of the Mana floodplains.
42. Interview, Bulawayo, June 19, 2003.
43. The British sometime-journalist Bernard Venables simply excises the lake from his descent of the river. "I bypassed Kariba," he explains very briefly, in favor of the "uninhabited, untouched bush" downstream (Venables 1974:197).

44. Interview, Stellenbosch, South Africa, May 18, 2004.
45. Regarding DDT see Douthwaite (1992) and Mhanga, Taylor, and Phelps (1986).
46. Interview, Harare, August 5, 2005.
47. In the same, contradictory vein, *Wet Breams*, the boozy, baudy memoir of Bill Taylor, introduces Mana Pools as "a flood plain before the Kariba Dam wall was built." On the next page, Taylor relaxes "in my deck chair on the bank just soaking it all up . . . 'God's Garden' " (B. Taylor 2002:147, 148).
48. St. Leger (2004:130). The phrase "maddening crowd" would seem to represent a corruption of the title of Thomas Hardy's novel *Far from the Madding Crowd* (1874), wherein the word "madding" carries quite a different meaning.
49. Interview, Harare, August 5, 2005.
50. Interview, Marondera, May 31, 2007.
51. Wilson (1991). The dichotomy is equivalent to Lefebvre's (1991) more famous distinction between "landscapes of production" and "landscapes of consumption."
52. Langston (2003); McPhee (1971:196ff). Further studies along these lines include Fiege (1999) and Swyngedouw (1999).
53. In addition to John Gordon Davis (see above), Duncan Watt, who was born in Zambia but has apparently resided in Singapore since 1980, writes of a plot to explode the dam wall (Watt 1992).

CHAPTER 3

1. Such logics of colonization, of course, suffered from multiple contradictions: Indians did cultivate, although not on permanent, fenced, intensive plots, while colonists relied upon the labor of other groups, notably African slaves.
2. For analysis of state-led violence in this period, see Duffy (2000).
3. Interview, Hoedspruit, South Africa, February 25, 2006.
4. McGregor (2003:94) confirms that, although they may understand what a river is, the Tonga, Leya, or Dombe people do not use any proper name for the Zambezi.
5. He did, nonetheless, write and disseminate a long defense of Mozambique Safaris (Edwards 1994).
6. Interview, Hoedspruit, South Africa, February 25, 2006.
7. Lemon (1987) first published a guidebook to Kariba. Then, he released a children's version of his adventure tale in 1988. Used in schools, it was reprinted in 1992, 1994, 1995, 1996, 1999 (three times), and 2000. The adult version of his voyage appeared

in 1997. Finally, having taken up part-year residence in Britain, he printed his own memoirs of his service in the police force (2000).

8. Lemon (1997:17). Veronica Stutchbury (1992:17) also mentions "unexplored bays of the Ume River" where it empties into Lake Kariba.
9. Interview, Heathrow, UK, August 2, 2003.
10. Lemon (1997:87, 112–113). The children's version (1988:32) includes the first phrase verbatim.
11. Interview, Heathrow, UK, August 2, 2003.
12. Interview, Heathrow, UK, August 2, 2003.
13. Interview, Redwood City, California, March 11, 2007.
14. Interview, Redwood City, California, March 11, 2007.
15. McGregor (2003) describes the practices, beliefs, and political claims of inhabitants along this section of river.
16. An earlier generation of foreign-born Kariba writers had performed a similar move, encouraging readers to put the dam out of their minds (see Chapter 2).
17. Interview, Gaborone, February 26, 2006.
18. Interview with Rex Taylor, Kariba, July 2, 2003. According to Taylor, one place name, Kennies Island, does refer to African nationalism: to Kenneth Kaunda, the independence leader of Zambia, who liked to vacation there.
19. See M. Davis (1998:11–14) and Carter (1987:148–149). Campbell (1988:22) and Raffles (2002:101) discuss similar borrowing of the terms for geographical features in other colonial locales. North Americans seem to have assimilated non-English proper names more easily. See the Preface's anecdote regarding the different levels of anxiety provoked by "Massachusetts" and "Chimanimani" among relevant white populations.
20. Pitman (n.d.:52). This book, *Zimbabwe Portrait*, sold roughly 10,000 copies as well.
21. Interview, Harare, June 30, 2003.
22. Interview, by telephone, June 7, 2004.
23. Interview, Harare, July 28, 2003
24. Interview, Harare, July 28, 2003.
25. Interview, Kariba, July 2, 2003.
26. Chiefs Mola, Negande, and Sampakaruma.
27. Interview, Victoria Falls, March 10, 2005.
28. Interview, Kariba, May 27, 2007.
29. Kennedy Matarisagungwa is a pseudonym.
30. Interview, Kariba, May 27, 2007.

31. Patrons appear to more readily associate the hotel with the Scotch whisky named Cutty Sark, once consumed in large quantities there.
32. Interview, Kariba, May 28, 2007. See Behrend (2000:73) for a similar use of images to "give presence to . . . a richer, cosmpolitan world . . ."
33. Brian Keel is a pseudonym.
34. In 2006 or 2007, the (white) mayor of Kariba proposed the same relocation for the same reasons (Houghton n.d.).
35. Interview, Harare, May 28, 2007.
36. According to Kenmuir, the novel was runner-up for the Young Africa Award of 1993. Interview, Fishhoek, South Africa, May 17, 2004.
37. *The Fisherman*'s two videos on KITFT make this lopsided participation apparent (*Zimbabwe Fisherman* 1999; *African Fisherman* 2002). "This particular tournament has such historical connotation with the country, with the culture, with its people," avers a white male angler interviewed in the 1999 video. See Butson (1993), Jubb (1961:79–80), and Kenmuir (1983:58–62) for the most authoritative descriptions of the tiger fish and of techniques for catching it. Ironically, the species only habituated to the open water of the lake from 1970 onward in response to the introduction of kapenta sardines from Lake Tanganyika.
38. Interview, Fishhoek, South Africa, May 17, 2004.
39. The Harare-based firm Ink Spot designed the calendar in 1993. Staff of IUCN wrote the text first, and then the director of Ink Spot, Paul Wade, suggested using a Landsat TM image and found the image eventually used (Paul Wade, interview, by telephone, April 20, 2007).
40. Interview, Vumba, Zimbabwe, May 24, 2007.
41. Perhaps as a means of offsetting this exclusion, Paul Wade added a silhouette of a man and woman along one side of the satellite image.
42. A. Williams (2001a:43), excerpting a report compiled by Johnny Rodrigues.
43. Glen Powell is a pseudonym.
44. Interview, Harare, November 19, 2002.
45. Interview, Harare, November 19, 2002.
46. After 2000, wealthier whites unmoored themselves from the hotels altogether and took their holidays by rented houseboat.
47. "Wildlife industry loses 70 percent of its animals," *The Financial Gazette*, March 27, 2003, p. 31. The article relies upon a report by Wildlife and Environment Zimbabwe, the successor to the Wildlife

Association of Zimbabwe. See also Jenny Sharma, "ZANU-PF's legacy to Zimbabwe," unpublished report, November 2001.

48. Wynn (1998:39ff; 2000) provides an analysis of the organization's survey of its members and tourists.
49. The Communal Areas Management Programme for Indigenous Resources (CAMPFIRE) represented the height of this quid pro quo. For a critical review, see Hughes (2001, 2005).
50. For an initiative of this sort, see the "landshapes" and "manshapes" images of the late South African aerial photographer Herman Potgieter (1990:esp. 54–57).

CHAPTER 4

1. Kennedy (1987:2–3). The white population crested in roughly 1975 at 278,000, as against nearly 7 million blacks (Godwin and Hancock 1993:287).
2. Despite missing its targets by a wide margin, the land reform program, on balance, benefited the 70,000 resettled families substantially (Kinsey 1999).
3. A letter to the editor, for example, argues, "There were no black landowners to steal it [land] from then [before 1900]. Local blacks . . . preferred to live a nomadic life." The anonymous author described him- or herself as "3rd generation 'white Zimbabwean,' 9th generation 'white African,' hence Afrikaner" (*The Daily News* [Harare], September 24, 2002, p. 7). For a discussion of the ways in which white pioneers deliberately settled in proximity to resident blacks—whom they needed for labor—see Hughes (2006:54–55).
4. Appadurai (1991); cf. Clifford (1997:17–46). Rutherford (2001:80–81) describes commercial farmers in northern Zimbabwe in this fashion.
5. Godwin and Hancock (1993:287); Buckle (2002:63). Note: my subjects frequently described themselves as "patriots," never as "nationalists."
6. In this connection, Uusihakala (1999:39) refers to "double diaspora" of white Kenyans (cf. Ward 1989:1; Wagner 1994:7).
7. In referring to "Virginia," this work follows the boundaries of the intensive conservation area bearing that name. Most farmers distinguished between the eastern, lower-elevation side of this area, Virginia proper, and the western, higher side, Macheke.
8. Clements and Harben (1962:28–33) summarize this history. On the cultural meanings of Raleigh, Elizabeth, and Virginia, see Lim

(1998). Perhaps not unrelated to this link between the Virginias, a settler arriving in Marondera shortly after WWII named his farm "Raleigh" explicitly after Sir Walter Raleigh (English 1995:81).

9. Clements and Harben (1962:27). Edward Harben was vice president of the Rhodesian Tobacco Association, an industry group of growers, from 1946 to 1954 (cf. Rubert 1998).
10. The importance of the watershed to whites' view of the eastern Mashonaland landscape can hardly be overemphasized. The front cover of the 1972 agricultural survey of Marondera shows an aerial photograph with the watershed lines added in (Ivy and Bromley 1972). In 1987, a group of whites founded Watershed College in Wedza, slightly to the south of Marondera (Bissett 2003:45). English's (1995) reminiscences of life in Wedza from the 1920s to the 1940s refers repeatedly to the watershed.
11. British South Africa Company, *Information for Intending Settlers*, 1901, p. 34; cited in Kennedy (1987:121).
12. Interview, Marondera, October 1, 2002.
13. See Meadows (1996) and Nyschens (1997) for frequent uses of this phrase.
14. Quoted in R. Reynolds (1964:2). Hodder-Williams (1983:45–68) and MacDonald (2003:53) provide fuller accounts of Scorror and early settlement.
15. Interview, Virginia, November 14, 2002.
16. Interview, Marondera, January 16, 2002.
17. Interview, Marondera, January 17, 2002. Cf. Foster (2008:262).
18. Regarding further transfers of soil science from the United States, see Anderson (1984) and Dodson (2004).
19. See Hodder-Williams (1983:199) regarding early ICAs. The terraces work best on slopes of 6–8 percent and comprise a cut trough and filled ridge. They are distinguished from the more well-known bench terraces, which apply to much steeper slopes and resemble a flight of stairs (Schwab et al. 1993:154–155). Zimbabwean farmers refer to the broadbase terraces colloquially as "contour ridges" (cf. Elwell n.d.)
20. By the 1970s, the Rhodesian Front government was sponsoring a set of conservation organizations internal to the tribal trust lands. These did not succeed either. Regarding the failure of such conservation protocols among Africans, see Drinkwater (1991), Hughes (2006:67), McGregor (1991), Weinrich (1975:151), and K. Wilson (1989).

21. In the same spirit, D. S. McClymont's (1981) 90-page review of tobacco advice mentions labor only five times. The Agricultural Labor Bureau is a committee of the Commercial Farmers' Union.
22. Measured between the 1,100 and 1,500 m contour lines on the 1:50,000-scale Macheke, St. Benedict Mission, Munda maps (respectively, numbers 1732C3, 1832A1, and 1831B2 from the Zimbabwe Office of the Surveyor General). This gradient of 1.13 percent actually exceeds the recommended slope for parallel layouts (see below).
23. Interview, Harare, November 22, 2002.
24. Ostrom (1990:91–92). Cited by Virginia whites, the pre-2000 figure of 72 families is probably understated. Throughout, I am using the terms "farm" and "family" loosely. Large family units frequently managed multiple, adjacent farms jointly and/or as a corporation. The issue of arable land was hotly contested throughout the 1990s. The state frequently claimed that farmers failed to use their land fully. Farmers suggested that plowing marginal areas would ultimately wreck the soil.
25. Roughly ten farmers attended annual general meetings.
26. Interview, Harare, March 14, 2003.
27. To represent communal land residents, the Virginia ICA invited a staff member of Agritex, the agricultural extension agency, who (after independence) would have been black.
28. Interview, Harare, July 18, 2003.
29. Interview, Harare, March 14, 2003.
30. Although local agents of the ruling party and/or the secret police carried out the attack, the involvement of party officials in Harare remains unclear.
31. For the quotation: interview, Harare, June 10, 2003. Regarding Stevens's career, see Buckle (2002:53–54).
32. Interview, Harare, May 26, 2003 (cf. Fortmann 1995:1058–1059).
33. Up to 1991, the Virginia ICA hired a government plane. When that aircraft became unavailable, Gemmill volunteered his own (Virginia ICA, minutes of annual general meeting held on September 26, 1991, p. 2).
34. Interview, Harare, June 10, 2003.
35. Interview, Harare, November 22, 2002.
36. Virginia ICA, minutes of meeting held on June 6, 1996, p. 2.
37. Interview, Harare, May 26, 2003.

38. Interview, Harare, March 14, 2003.
39. Interview, Marondera, July 30, 2002.
40. Interview, Virginia, November 14, 2002.
41. Interview, Virginia, November 7, 2002.
42. Virginia ICA, minutes of the annual general meeting, September 26, 1991.
43. Interview, Harare, October 29, 2002.
44. Interview, Virginia, November 14, 2002.
45. Wood (2003) provides a full analysis of dam construction in Virginia.
46. On the more arid lowveld, inadequate runoff greatly reduced the potential for dam building. These areas, in any case, lay outside the symbolic and political heart of white agriculture.
47. Anonymous posting to the Justice for Agriculture (Harare) "Open Letter Forum" email list, entitled "A return to Macheke/Virginia farming area," May 19, 2003.
48. After nationalizing derelict and abandoned farms between 1980 and 1983, the state, in fact, acquired very little land. Owners wishing to sell farms had to petition the state for a "certificate of no interest," and the state almost always granted it. A considerable portion of Zimbabwe's commercial farmland changed hands—almost entirely between whites (Rugube et al. 2003:129).
49. Act 3/1992, the Land Acquisition Act. Cf. Moyo (2000:75).
50. Interview, Ruzawi, July 23, 2002.
51. Interview, Marondera, October 10, 2002.
52. Interview, Harare, May 26, 2003.
53. Interview, Marondera, May 1, 2003.
54. Interview, Virginia, July 23, 2003.
55. Interview, Harare, June 10, 2003. Tom Sweeney is a pseudonym. Sweeney overheard the remark in Shona—apparently made on the assumption that he did not understand that language—and recounted it to me and others in English.
56. Interview, Virginia, May 22, 2003. Constantine Gavras is a pseudonym.
57. Regarding the vastly larger Lake Kariba, Soils Incorporated (2000:73) confirms only a cooling effect "in the immediate vicinity of the lake." In any case, wind would carry evaporated water a considerable distance before it precipitated.
58. Interview, Harare, July 18, 2003.
59. Interview, Virginia, July 23, 2003. Henk Jelsma is a pseudonym.

60. The relationship is slightly more complex because dams tend to grow wider as they grow taller, dimishing seepage in a linear fashion (Schwab et al. 1993:197–201).
61. Interview, Marondera, June 12, 2003. Don Lanclos is a pseudonym. Since spillway water is free of sediment (the sediment having fallen in the still water of the reservoir), it has a high capacity to pick up sediment as it accelerates (McCully 1996:33).
62. Virginia ICA, minutes of meeting held on August 1, 1996, p. 1.
63. Virginia ICA, minutes of the 44th annual general meeting, September 17, 1997, p. 1.
64. Virginia ICA, minutes of meeting held on February 4, 1999, p. 2.
65. Interview, Harare, October 18, 2002.
66. This irrigation infrastructure differed fundamentally from the gravity-fed systems more typically studied by anthropologists (e.g., Geertz 1963; Lansing 1991).
67. Interview, Marondera, March 20, 2003. From the late 1990s, Wood had served as chairman of the Nyagui Sub-Catchment Council.
68. Interview, Macheke, July 31, 2002. Johann Swanepoel is a pseudonym.
69. Virginia ICA, minutes of the 40th annual general meeting, September 23, 1993, p. 1.
70. Virginia ICA, minutes of the 43rd annual general meeting, 1996, p. 1.
71. Interview, Marondera, July 30, 2002.
72. Interview, Harare, August 5, 2002.
73. Interview, Harare, March 14, 2003.
74. Interview, Marondera, March 20, 2003.
75. Interview, Marondera, May 1, 2003.
76. Interview, Harare, June 10, 2003.
77. Interview, Harare, May 26, 2003.
78. All shoreline lengths apply at reservoirs' full supply level. Les Wood estimated most of the shorelines using original builders' basin surveys (Wood 2003).
79. Interview, Marondera, May 1, 2003.
80. Chingezi is a pseudonym.
81. Interview, Marondera, July 23, 2003
82. The longest reservoirs' shorelines were 17.25, 15.15, and 13.13 km. Wood (2003) calculated two others at 10.50 and 10.05 km, but these lie within the measurement error of Chingezi's shoreline.
83. Interview, Virginia, July 23, 2003.

84. Interview, Harare, March 14, 2003.
85. John Tessmer is pseudonym, as is Shiri Farm.
86. Virginia ICA, minutes of the annual general meeting held September 17, 1997.
87. Interview, Marondera, June 12, 2003.
88. The African Waterfowl Census (carried out under the auspices of the International Waterfowl and Wetlands Research Bureau) conducted bird counts on one site in the vicinity of Virginia: Don's dam near Rusape. These findings show wild oscillations in the presence of Egyptian goose (*Alopochen aegyptiacus*) and knobnose or comb duck (*Sarkidiornis melanotos*) between 1993 and 1998. On the other hand, if the aggregate bird population were stable, one would expect decreasing bird densities during this period of major reservoir filling (as birds dispersed themselves to more and more habitats). The absence of a clear decline in bird populations at any one site could indicate an increase in aggregate populations visiting eastern Zimbabwe. I thank Peter Rockingham-Gill for making these raw data available to me.
89. Interview, Harare, July 3, 2003. Officially entitled the Zimbabwe National B.A.S.S. Federation, the Zimbabwe Bassmasters constituted a branch of B.A.S.S., the global Bass Anglers Sportsman Society, based in the United States.
90. Interview, Marondera, May 1, 2003.
91. In the late 1990s, various estates began labeling themselves as "safari farms" or "holiday farms" (Irene Staunton, personal communication, July 7, 2004; Mark Guizlo, personal communication, June 10, 2002; cf. Kramer 2003). I would distinguish this phenomenon from the slightly earlier conversion of large-scale cattle ranches into wildlife conservancies in the lowveld.
92. Frank Richards is a pseudonym.
93. Interview, Harare, March 17, 2003.
94. Interview, Marondera, June 12, 2003.
95. Regarding the association of "nature" and artificial water—particularly in the western United States—see Fiege (1999), Hughes (2005), Langston (2003), McPhee (1971), and White (1995).
96. Interview, Harare, June 10, 2003.
97. Interview, Harare, October 28, 2002.
98. Interview, Harare, October 29, 2002.
99. At the national level, however, commercial farmers had taken significant steps toward supporting an emergent class of black

landowners. The Zimbabwe Tobacco Association, a part of the Commercial Farmers' Union, systematically sponsored black tobacco growers through apprenticeships and supervised, independent farming.

100. See Davies (2001) for a treatment of this issue with respect to urban whites.
101. Interview, Virginia, May 21, 2003.

CHAPTER 5

1. White numbers plummetted from 232,000 in mid-1979 to roughly 80,000 in 1990 (Godwin and Hancock 1993:315).
2. Godwin and Hancock (1993:7) and Selby (2006:102–103) are at pains to demonstrate a diversity of political opinion among whites of the 1970s. Such variation, however, only embraced gradual, measured reform, falling short of support for any immediate revolution toward majority rule.
3. Judith Todd's recent memoir discusses her activity and that of Guy Clutton-Brock and John Conradie, admitting that she "never had as tough a time as the Slovos" (Todd 2007:12, 20–21, 23, 264; cf. Chung 2005:58). Some white missionaries also assisted the guerrillas covertly (Maxwell 1999:125–129). Jeremy Brickhill served in the intelligence corps of ZIPRA, the guerrilla army of the Zimbabwe African People's Union (ZAPU) (cf. Brickhill 1995).
4. Ian Smith, however, continued to criticize the government, as his memoirs attest (Smith 2001).
5. Herbst (1988/89:46–47) describes a slightly different "racial bargain" in which whites were allowed to retain their material wealth in return for the emigration of their children. Yet, as Herbst also explains, the state exempted commercial farmers from that obligation.
6. See Lessing (1992) for wry descriptions of "the monologue" she heard from Zimbabwean whites in the 1980s.
7. ZANU-PF stands for Zimbabwe African National Union – Patriotic Front.
8. The MDC won 57 of 120 contested seats. Regarding the involvement of white farmers, see Lamb (2006:199, 241–244) and Taylor (2002:10).
9. The term "war veterans" has been used loosely and includes many individuals who did not take part in the armed struggle of the 1970s. Among the numerous accounts of the state's motives and

of the historical ruptures and continuities regarding the invasions, see J. Alexander (2006:184ff), Cousins (2006:594–596), Hammar and Raftopoulos (2003), and Moyo (2005).

10. As exemplars of that research, see Rukuni (1994) and Moyo (1995). Selby (2006:208–209) summarizes debates on the land tax and the CFU's position therein in the 1990s. A high-level donors' conference held in 1998 generated still more text on beneficiary selection, subdivision, land taxes, and so on, including Hughes (2000) and, although overtaken by events, Michael Roth and Francis Gonese's (2003) massive contribution.
11. The phrase "anti-politics machine" comes from Ferguson (1990). See J. Alexander (2006) for a history of technocratic land use planning in Rhodesia and Zimbabwe.
12. In fact, Zimbabwe banned dual citizenship, but foreign embassies obstructed the law. Typically, an individual was required to submit the foreign passport when applying for a Zimbabwean passport. The Government of Zimbabwe would send that document to the high commission or embassy of the relevant country, which would immediately send it back to the holder.
13. Davies (2001:242). Writing shortly before Zimbabwe's independence, Marshall Murphree—himself a white Zimbabwean—makes the same point with respect to whites in other former colonies (Murphree 1978:169).
14. Most notably, Mugabe appointed Denis Norman, president of the Rhodesia National Farmers Union (the precursor of the CFU), as minister of agriculture in his first cabinet.
15. Anonymous, "A return to Macheke/Virginia farming area," Justice for Agriculture Open Letters Forum, no. 81, May 16, 2003.
16. Interview, Harare, June 10, 2003.
17. Interview, Virginia, May 21, 2003.
18. Dawn Harper is a pseudonym. Quoted in Staunton (2005:467).
19. Buckle (2001:50ff) and Godwin (2006:63–64) also provide accounts of the murder and subsequent events.
20. Staunton (2005:472). See Lamb (2006:202) for similar views among Stevens's associates.
21. Interview, Macheke, July 31, 2002. See Holtzclaw (2004:217) regarding a similar discussion between a farm owner and farm workers.
22. The matter, in any case, was overdetermined. Corruption, repression, and the violence concurrently directed against them gave farm workers ample reason to reject ZANU-PF (Rutherford 2001:246–247).

23. Commercial Farmers' Union, Farm Invasions Update, August 22, 2000. See Wiles (2005:22) for a similar example of war veterans' actions to "demean and lower the standing of [farm owners] . . . in the eyes of the people we employ."
24. Interview, Virginia, August 6, 2005. See Buckle (2001:76).
25. Commercial Farmers' Union, Farm Invasions and Security Report, March 8, 2002. Cf. Selby (2006:291).
26. Interview, Macheke, July 31, 2002.
27. Commercial Farmers' Union, Farm Invasions and Security Update, November 10, 2000.
28. Indeed, in 2005, the prominent ZANU-PF politician Fay Chung wrote, "Most people . . . do not see anything wrong with the confiscation of white-owned property" (Chung 2005:328).
29. Interview, Marondera, July 9, 2002.
30. AIG Zimbabwe. No date. "Famers Comprehensive Policy." Harare: AIG Zimbabwe, p. 7.
31. Interview, Harare, May 16, 2006.
32. Muzondidya (2007:33). Raftopoulos (2005:xiii) and Ranger (2004) also examine the content and meaning of state propaganda. In this sense, Hellum and Derman minimize the importance of race and racism when they write, "Fast track . . . is about the recentralization of power and resources" (Hellum and Derman 2004:1799).
33. Herein lay the rhetorical power of a symbolic surrender of property. In 2000, Eastern Highlands coffee grower Roy Bennett even told the press, "They can take my farm, but I am still running" for Parliament. Once elected, however, he did not relinquish his estate until war veterans drove him from it by force. Ross Herbert, "White Zimbabwean takes a stand," *Christian Science Monitor*, May 17, 2000. The state persecuted Bennett continuously thereafter.
34. Presentation by Jenni Williams to the Civic Leaders Consultative Meeting, convened by Justice for Agriculture, Meikles Hotel, Harare, August 6, 2002.
35. Steve Pratt. 1998. "Kudu Drift: a changing landscape."
36. Interview, Marondera, July 12, 2002.
37. Interview, Totnes, UK, March 1, 2005.
38. Quoted in Buckle (2001:28). In print, Doris Lessing (2003:10) and the ex-farmer Richard Wiles (2005:38) also dwell on the negligent and/or intentionally cruel manner in which certain black occupiers treated animals. See Shutt (2002:272–274) for a history of white etiquette toward livestock.

39. Interview, Marondera, January 17, 2002.
40. Frank Richards is a pseudonym. Interview, Harare, October 28, 2002.
41. Commercial Farmers' Union, Farm Invasions and Security Reports, July 12, 2001 and September 20, 2001.
42. Commercial Farmers' Union, Farm Invasions and Security Reports, December 3, 2001, February 25, 2002, March 15, 2002.
43. Interview, Harare, October 28, 2002.
44. Gumbo (1995:197, 182). In his only work of historical fiction, Daneel writes under a Shona penname. His other work centers on the independent, Zionist churches in Zimbabwe, in whose unity and organization he has also played a leadership role.
45. Stevenson is a pseudonym.
46. Farmers for Jesus. No date. "Well-watered gardens: instructions for the planting of a plot." Harare: Farmers for Jesus.
47. Interview, Harare, May 15, 2006.
48. Interview, Marondera, October 1, 2002. The quotation paraphrases a passage from 1 Kings 21:3.
49. Another Virginia family invited a traveling pastor of the conservative, South Africa – based Rhema Bible College to evangelize among the workers. As born-again Christians, these individuals showed greater affinity with such right-wing churches (Gifford 1988) than with the social gospel of the mainline Protestant denominations (Bornstein 2003).
50. Paul and Mary Fisher are pseudonyms. Interview, Virginia, May 22, 2003.
51. Interview, Virginia, May 30, 2007. I have changed the name in the quotation from the original to "Fisher."
52. Interview, Virginia, May 30, 2007.
53. Letter from M. E. Nyamukapa, Chief Executive Officer, Murehwa Rural District Council, June 13, 2003.
54. Interview, Virginia, August 7, 2005.
55. In this connection, Ivan Evans refers to the "bastardization of authority" (I. Evans 1997:190; cf. Hughes 2006:152–153).
56. Interview, Virginia, November 14, 2002.
57. Interview, Harare, November 29, 2002.
58. Interview, Virginia, May 22, 2003.
59. Interview, Harare, March 17, 2003. See Worby (2003:75–76) regarding the ambivalence of many commercial farmers to this plan.

60. Statutory Instrument 346 of 2001, officially known as the Land Acquisition (Offers of Land in Substitution for Land to be Acquired for Resettlement Purposes and Related Matters) Regulations. The decree invited farmers to submit "LA 3" forms proposing reduced parcels (Selby 2006:289). In Manicaland and Midlands—as opposed to all other provinces—government appears to have implemented this legislation and allowed owners to downsize their farms (Selby 2006:322).
61. Interview, Austin, Texas, USA, June 28, 2006.
62. Interview, Ruwa, August 9, 2005.
63. Interview, Marondera, May 31, 2007.
64. His son—also displaced—bought the farm and acquired the skill of sheep husbandry.
65. Stanley Hayes is a pseudonym. Interview, Acton, Ontario, Canada, July 7, 2005. In the first quotation, I have replaced the original name with that of Hayes.
66. From [original name deleted] & Son (Pvt) Ltd. to whom it may concern, June 10, 2001.
67. Interview, Murehwa, May 26, 2007.
68. Interview, Acton, Ontario, Canada, July 7, 2005.
69. Although no verifiable figures exist, Selby (2006:319–320) estimates that roughly 600 families were still farming in 2004, of which 200 resided off-farm.
70. Two of these 11 were working as managers on white-owned farms and leasing in some land to farm on their own. Before 2000, a portion of the original 75 families was similarly situated, but it is impossible to know how many families at this point.
71. Officially known as the Labour Relations (Terminal Benefits and Entitlements of Agricultural Employees Affected by Compulsory Acquisition) Regulations, 2002.
72. Selby (2006:307–308). Rutherford (2003:205–211) sheds much light on farm workers' complex strategies—of playing their employers and occupiers against each other—pursued elsewhere in Mashonaland East.
73. Peter Farnsworth is a pseudonym. Interview, Virginia, May 21, 2003.
74. According to scattered anecdotes, however, some chefs hired displaced white farm owners to manage farms owned by chefs.
75. These politicians, whose number was impossible to verify, took farms in Macheke.

76. The number of farm workers among these beneficiaries is much debated and nearly impossible to verify (Marongwe 2003b:19; Rutherford 2003:211).
77. During this period, government spokespeople continually referred to 99-year leases it would issue, but those documents never appeared.
78. Interview, Virginia, May 21, 2003. On similar arrangements, see Marongwe (2003a:183); and Selby (2006:302), who describes them as leading to farmers' eviction from Concession District, Mashonaland Central.
79. Interview, Murehwa, May 26, 2007. With stronger political ties than most A2 farmers, Nick Mangwende is a nephew of former minister and senator Witness Mangwende, and a son of the nephew of Chief Mangwende, resident in the adjoining Mangwende Communal Area.
80. Justice for Agriculture, Open Letter Forum, October 15, 2002.
81. Walter Finch is a pseudonym. Interview, Harare, March 14, 2003.
82. Notes from a Farmers' Forum meeting, Virginia, November 14, 2002.
83. Henry Hart and Roy Baker are pseudonyms. Notes from a Farmers' Forum meeting, Virginia, June 11, 2003.
84. Interview, Macheke, July 21, 2002.
85. Remarks by Dave Conelly and Jenni Williams, respectively, at a "sundowner presentation" convened by Justice for Agriculture, Chapman Golf Club, Harare, August 1, 2002.
86. Notes from Civic Leaders Consultative Meeting convened by Justice for Agriculture, Meikles Hotel, Harare, August 6, 2002.
87. Interview, Harare, August 4, 2005. Worswick, the chair of JAG, also served as its press spokesperson, the function Williams had performed.
88. "Land reform goes on trial," *The Zimbabwe Independent*, Harare, November 3, 2006.
89. In April 2009, the International Center for Settlement of Investment Disputes awarded compensation to the litigants, none of which had been paid as of October 2009. Meanwhile, a tribunal of the Southern African Development Community (SADC) rendered a number of judgments in favor of Zimbabwean commercial farmers. Rather than implementing these rulings, the government of Zimbabwe arrested and beat some the most prominent litigants.
90. Interview, Macheke, September 26, 2002.
91. Interview, Virginia, May 21, 2003.
92. Interview, Macheke, July 31, 2002.

93. Notes from Macheke farmers' meeting, Macheke, October 10, 2002.
94. Letter from Bruce Gemmill to Justice for Agriculture Open Letter Forum No. 382, August 19, 2005. The convoluted struggle resulted in the creation of the Justice for Agriculture Membership Association, on whose board Gemmill served.
95. Interview, Virginia, May 31, 2007.
96. Interview, Virginia, August 8, 2005.
97. Interview, Virginia, May 20, 2007.
98. Letter from Chief Mangwende to the district administrator, Murewa, March 18, 2006.
99. Moyo (2000:90ff) describes a shift among commercial farmers away from more extensive, regulated, and taxed crops, such as maize and tobacco, and toward specialty production, especially horticulture.
100. Notes from meeting of the Virginia A2 Farmers' Association, Macheke Club, Macheke, May 26, 2007.
101. Virginia, August 6, 2005.
102. Lisa Farnsworth is a pseudonym. Interview, Virginia, May 29, 2007.
103. "A Farm in Zim," May 8, 2007. Emphasis added.
104. Interview, Virginia, August 6, 2005.
105. Outside government circles, it was widely accepted that ZANU-PF won the 2000 and 2002 elections through fraud and voter suppression.
106. Interview, Virginia, August 7, 2005. Written by a black Zimbabwean resident in the United States, Chris Gande's (2005) novel—wherein a white farmer's daughter falls in love with the son of a government minister—could not have been further from the truth.
107. Interview, Harare, November 29, 2002.
108. Interview, Ruzawi, October 10, 2002.
109. Interview, Murehwa, May 26, 2007.
110. Held at the Centre for Rural Development, University of Zimbabwe, Harare, August 5, 2005. On the other hand, for an unusually positive assessment of the new economic relations in agriculture, see Mavedzenge et al. (2008).

CHAPTER 6

1. Lee Gavras and Constantine Gavras (see below) are pseudonyms.
2. This dialogue took place in Greek as follows. Lee: Λεγετε, μιλαγε. Constantine: Αυτο το φραγμα το φτιαξαμε να αφισουμε

και εμεις κατι στους μαυρους. Τοσα χρονια ζουμε στιν Αφρικη και μας περιπιουντα οι ανθροποι. Lee: Και τι ενοιες? Constantine: Να κατσις στο μερος σου. I am grateful to Pano Yannakogeorgos for this translation.

3. Louter (2006:127). Louter quotes from A. Starker Leopold et al., "Wildlife management in the National Parks," March 4, 1963, reprinted in Department of the Interior, *Administrative Policies for Natural Areas of the National Park System* (Washington, DC: Government Printing Office, 1970), p. 104.
4. Rogers (2005:30, 50, 53, 60–61, 75, 122–123, 136–137) presents advertising images of the same.
5. The South African novelist Zakes Mda (2000:109–110, 117) alludes to the debates on this form of tourism.
6. Regarding "second nature" and "nature-society hybrids," see Pollan (1991) and Zimmerer (2000), respectively.
7. Launched by MacArthur and Wilson (1967), the field of island biogeography addresses such questions of scale, minimum habitat sizes, and diversity.
8. Cosgove (1984:268) describes a similar and earlier movement of walkers in Britain.
9. Robinson's text and the images appear on http://www.spirit-of-the-land.com/galleries.htm. Downloaded on March 1, 2007. I interviewed Robinson in Livingstone on July 7, 2004.
10. Unconfirmed reports suggest, however, that the frequent passage of rafts has discouraged local residents from smuggling goods across the river.
11. Jonathan Adams and Thomas McShane (1992) and Joubert (2006) are notable exceptions.
12. Coetzee (1999:158–159, 202–204). See Rita Barnard's (2007: 38–40) similar analysis of the same passages.
13. Among black Zambians, Ferguson (1999) describes as "abjection" a similar mentality of frustrated expectation.

References

Abbey, Edward. 1968. *Desert Solitaire: A Season in the Wilderness.* New York: McGraw-Hill.

Adams, Jonathan S., and Thomas O. McShane. 1992. *The Myth of Wild Africa: Conservation without Illusion.* Berkeley: University of California Press.

African Fisherman. 2002. "Kariba International Tiger Fishing Tournament" (video). Harare: African Fisherman.

Alexander, Jocelyn. 2006. *The Unsettled Land: State-Making and the Politics of Land in Zimbabwe, 1893–2003.* Oxford, UK: James Currey.

Alexander, Karin. 2005. "Orphans of the empire: an analysis of elements of white identity and ideology construction in Zimbabwe." In Raftopoulos and Savage, eds., pp. 193–212.

Allina, Eric. 1997. " 'Fallacious mirrors': colonial anxiety and images of African labor in Mozambique, ca. 1929." *History in Africa* 24:9–52.

Anonymous. 1959. *Lake Kariba: The Story of the World's Biggest Man-Made Lake.* Bloemfontein, South Africa: The Friend Newspapers, Ltd. and the Central News Agency, Ltd.

Anderson, David. 1984. "Depression, Dust Bowl, demography and drought: the colonial state and soil conservation in East Africa during the 1930s." *African Affairs* 83(332):321–343.

Appadurai, Arjun. 1991. "Global ethnoscapes: notes and queries for a transnational anthropology." In Robert Fox, ed., *Recapturing Anthropology.* Santa Fe, NM: School of American Research Press, pp. 191–210.

Appiah, Kwame Anthony. 2006. *Cosmopolitanism: Ethics in a World of Strangers.* New York: Norton.

Armstrong, Peter. 1979. *Operation Zambezi: The Raid into Zambia.* Salisbury: Welston Press.

Ashcroft, Bill, Gareth Griffiths, and Helen Tiffen. 2002 [1989]. *The Empire Writes Back: Theory and Practice in Post-Colonial Literature.* 2nd ed. London: Routledge.

Ballinger, W. A. 1966. *Call it Rhodesia*. London: Mayflower.

Balneaves, Elizabeth. 1963. *Elephant Valley: The Adventures of J. McGregor Brooks, Game and Tsetse Officer, Kariba*. New York: Rand McNally.

Balon, E. K., and A. G. Coche, eds. 1974. *Lake Kariba: A Man-Made Ecosystem in Central Africa*. The Hague: Junk.

Barnard, Rita. 2007. *Apartheid and Beyond: South African Writers and the Politics of Place*. Oxford, UK: Oxford University Press.

Beach, D. N. 1980. *The Shona and Their Neighbors, 900–1850*. Gweru, Zimbabwe: Mambo Press and London: Heinemann.

Behrend, Heike. 2000. "'Feeling global': the Likoni Ferry photographers." *African Arts* 33(3):70–79.

Bell, Morag. 1993. "'The pestilence that walketh in darkness': imperial health, gender and images of South Africa." *Transactions of the Institute of British Geographers* new series, 18(3):327–341.

Bestall, Cliff. 2007. "Introduction, from a distance." In Kirkman, ed., pp. 6–7.

Bissett, Dave. 1995. "Watershed college." In Sheila Macdonald, ed., pp. 45–46.

Blackbourn, David. 2006. *The Conquest of Nature: Water, Landscape, and the Making of Modern Germany*. London: Jonathan Cape.

Bolze, Louis W. 1982. "Publisher's introduction." In *Four Voices: Poetry from Zimbabwe. By Rowland Molony, David Wright, John Eppel, and Noel Brettell*, Bulawayo, Zimbabwe: Books of Zimbabwe, pp. ix–x.

Bornstein, Erica. 2003. *The Spirit of Development: Protestant NGOs, Morality and Economics in Zimbabwe*. New York: Routledge.

Bourdillon, M. F. C., A. P. Cheater, and M. W. Murphree. 1985. *Studies of Fishing on Lake Kariba*. Gweru, Zimbabwe: Mambo Press.

Braudel, Fernand. 1980. *On History*. Chicago: University of Chicago Press.

Braun, Bruce. 2002. *The Intemperate Rainforest: Nature, Culture, and Power on Canada's West Coast*. Minneapolis: University of Minnesota Press.

Braun, Bruce. 2003. "'On the Raggedy Edge of Risk': articulations of race and nature after biology." In Moore, Kosek, and Pandian, eds., pp. 175–203.

Breytenbach, Breyten. 1996. *The Memory of Birds in Times of Revolution: Essays on Africa*. New York: Harcourt Brace.

Brickhill, Jeremy. 1995. "Daring to storm the heavens: the military strategy of ZAPU, 1976–79." In Ngwabi Bhebhe and Terence Ranger, eds., *Soldiers in Zimbabwe's Liberation War*. London: James Currey, pp. 48–72.

Buck, Pearl S. 1931. *The Good Earth*. New York: Grosset and Dunlap.

Buckle, Catherine. 1999. *Litany Bird*. Harare: College Press.

Buckle, Catherine. 2001. *African Tears: The Zimbabwe Land Invasions*. Johannesburg: Covos Day.

Buckle, Catherine. 2002. *Beyond Tears: Zimbabwe's Tragedy*. Johannesburg: Jonathan Ball.

Bunn, David. 2003. "An unnatural state: tourism, water and wildlife photography in the early Kruger National Park." In William Beinart and JoAnn McGregor, eds., *Social History and African Environments*. Oxford, UK: James Currey, pp. 199–220.

Burnett, D. Graham. 2000. *Masters of All They Surveyed: Exploration, Geography, and a British El Dorado*. Chicago: University of Chicago Press.

Burns, J. M. 2002. *Flickering Shadows: Cinema and Identity in Colonial Zimbabwe*. Athens, OH: Ohio University Press.

Butson, Bob. 1993. "Tigerfish." In Walsh and Williams, eds., pp. 1–5.

Campbell, Mary B. 1988. *The Witness and the Other World: Exotic European Travel Writing, 400–1600*. Ithaca, NY: Cornell University Press.

Carruthers, Jane. 1995. *The Kruger National Park: A Social and Political History*. Pietermaritzburg, South Africa: University of Natal Press.

Carter, Paul. 1987. *The Road to Botany Bay: An Exploration of Landscape and History*. New York: Knopf.

Carver, Maurice, and Robert Grinham. n.d. [ca. 1968]. *Ruzawi: The Founding of a School*. Salisbury: Ruzawi Old Boys Association.

Catholic Commission for Justice and Peace in Zimbabwe and Legal Resources Foundation. 1997. *Breaking the Silence, Building True Peace: A Report on the Disturbances in Matabeleland and the Midlands, 1980–1988*. Harare: CCJP and LRF.

Caute, David. 1983. *Under the Skin: The Death of White Rhodesia*. Evanston, IL: Northwestern University Press.

Chaumba, Joseph, Ian Scoones, and William Wolmer. 2003. "From *jambanja* to planning: the reassertion of technocracy in land reform in south-eastern Zimbabwe?" *Journal of Modern African Studies* 41(4):533–554.

Chennells, Anthony. 1982. "Settler myths and the Southern Rhodesian novel." Ph.D. dissertation, University of Zimbabwe, Harare.

Chennells, Anthony. 1995. "Rhodesian Discourse, Rhodesian novels and the Zimbabwe liberation war." In N. Bhebhe and T. Ranger, eds., *Society in Zimbabwe's Liberation War*. Vol. 2. Harare: University of Zimbabwe Press, pp. 102–129.

Chennells, Anthony. 2005. "Self-representation and national memory: white autobiographies in Zimbabwe." In Muponde and Primorac, eds., pp. 131–144.

Chung, Fay. 2005. *Re-Living the Second Chimurenga: Memories from Zimbabwe's Liberation Struggle.* Uppsala, Sweden: Nordic Africa Institute.

Clark, Nigel. 2005. "Postcolonial natures." *Antipode* 37(2):364–368.

Clements, Frank. 1959. *Kariba: The Struggle with the River God.* London: Methuen.

Clements, Frank, and Edward Harben. 1962. *Leaf of Gold: The Story of Rhodesian Tobacco.* London: Methuen.

Clifford, James. 1997. *Routes: Travel and Translation in the Late Twentieth Century.* Cambridge, MA: Harvard University Press.

Coetzee, J. M. 1988. *White Writing: On the Culture of Letters in South Africa.* New Haven, CT: Yale University Press.

Coetzee, J. M. 1999. *Disgrace.* New York: Penguin.

Cole, Monica M. 1960. "The Kariba project." *Geography* 45(1–2): 98–105.

Cole, Monica M. 1962. "The Rhodesian economy in transition and the role of Kariba." *Geography* 47(1):15–40.

Colson, Elizabeth. 1971. *The Social Consequences of Resettlement.* Manchester: Manchester University Press.

Comaroff, Jean, and John L. Comaroff. 2001. "Naturing the nation: Aliens, apocalypse and the postcolonial state." *Journal of Southern African Studies* 27(3):627–651.

Comaroff, John L. 1989. "Images of empire, contests of conscience: models of colonial domination in South Africa." *American Ethnologist* 16(4):661–685.

Cooper, Frederick, and Ann Stoler. 1989. "Introduction: tensions of empire: colonial control and visions of rule." *American Ethnologist* 16(4):609–621

Coppinger, Mike, and Jumbo Williams. 1991. *Zambezi: River of Africa.* London: New Holland Publishers.

Cosgrove, Denis. 1984. *Social Formation and Symbolic Landscape.* Madison: University of Wisconsin Press.

Cosgrove, Denis. 1988. "The geometry of landscape: practical and speculative arts in sixteenth-century Venetian land territories." In Denis Cosgrove and Stephen Daniels, eds., *The Iconography of Landscape.* Cambridge, UK: Cambridge University Press, pp. 254–276.

Cosgrove, Denis. 2005. "Tropic and tropicality." In Felix Driver and Luciana Martins, eds., *Tropical Visions in an Age of Empire.* Chicago: University of Chicago Press, pp. 197–216.

Cousins, Ben. 2006. "Review essay: debating the politics of land occupations." *Journal of Agrarian Change* 6(4):584–597.

Crapanzano, Vincent. 1986. *Waiting: The Whites of South Africa*. New York: Vintage.

Crapanzano, Vincent. 1994. "Kevin: on the transfer of emotions." *American Anthropologist* 96(4):866–885.

Cronon, William. 1983. *Changes in the Land: Indians, Colonists, and the Ecology of New England*. New York: Hill and Wang.

Cronon, William. 1995. "The trouble with wilderness; or, getting back to the wrong nature." In Willliam Cronon, ed., *Uncommon Ground: Rethinking the Human Place in Nature*. New York: Norton, pp. 69–90.

Crosby, Alfred. 1986. *Ecological Imperialism: The Biological Expansion of Europe, 900–1900*. Cambridge: UK: Cambridge University Press.

Davies, Angela C. 2001. "From Rhodesian to Zimbabwean and back: white identity in an African context." Ph.D. dissertation, University of California, Berkeley.

Davis, John Gordon. 1967. *Hold My Hand I'm Dying*. London: Michael Joseph.

Davis, John Gordon. 1972. *Operation Rhino*. London: Michael Joseph.

Davis, John Gordon. 1984. *Seize the Reckless Wind*. Glasgow: Collins.

Davis, Mike. 1998. *Ecology of Fear: Los Angeles and the Imagination of Disaster*. New York: Vintage.

de Lassoe, H. 1908. "The Zambezi River (Victoria Falls-Chinde): a boat journey of exploration, 1903." *Proceedings of the Rhodesia Scientific Association* 8(1): 19–50.

de Woronin, U. G. 1976. *Zambezi Trails*. Salisbury: Regal Publishers.

Didion, Joan. 1979. *The White Album*. New York: Simon and Schuster.

Dinesen, Isak. 1937. *Out of Africa*. New York: Random House.

Dobb, Lis. 1992. "How can we conserve our leisure-pleasure?" *The Zimbabwe Fisherman* 4(4):33.

Dodson, Belinda. 2004. "Above politics?: soil conservation in 1940s South Africa." *South African Historical Journal* 50:49–64.

Dominy, Michèle D. 2001. *Calling the Station Home: Place and Identity in New Zealand's High Country*. Lanham, MD: Rowman and Littlefield.

Douthwaite, R. J. 1992. "Effects of DDT on the fish eagle *Haliaeetus vocifer* population of Lake Kariba in Zimbabwe." *Ibis* 134:250–258.

Drinkwater, Michael. 1991. *The State and Agrarian Change in Zimbabwe's Communal Areas*. New York: St. Martin's.

Duffy, Rosaleen. 2000. *Killing for Conservation: Wildlife Policy in Zimbabwe*. Bloomington: Indiana University Press.

Dyer, Richard. 1997. *White*. London: Routledge.

Dzingirai, Vupenyu. 2003. "The new scramble for the African countryside." *Development and Change* 34(2):243–263.

Earl, Lawrence. 1954. *Crocodile Fever: A True Story of Adventure*. London: Collins.

Eastwood, Clint. 1990. *White Hunter, Black Heart*. Film directed by Clint Eastwood and produced by Warner Brothers.

Eco, Umberto. 1986. *Travels in Hyperreality*. San Diego, CA: Harcourt.

Ederg, Rolf. 1977. *The Dream of Kilimanjaro*. Translated by Keith Bradfield. New York: Pantheon.

Edwards, Stephen. 1974. *Zambezi Odyssey: A Record of Adventure on a Great River of Africa*. Cape Town: T.V. Bulpin.

Edwards, Stephen. 1994. "Wildlife, people and poaching: Cahora Bassa safari region, Mozambique." Unpublished document.

Elkins, Caroline. 2005. *Imperial Reckoning: The Untold Story of Britain's Gulag in Kenya*. New York: Henry Holt.

Elwell, H. A. n.d. [ca. 1975]. "Contour layout design." Harare: Department of Conservation and Extension.

English, Pat. 1995. *Lushington: A Fragment of Time*. Harare: Pat English.

Eppel, John. 2002. *The Holy Innocents*. Bulawayo: 'amaBooks.

Eppel, John. 2005. *Songs My Country Taught Me*. Harare: Weaver Press.

Eppel, John. 2006. *Hatchings*. Bulawayo: 'amaBooks.

Evans, Ivan. 1997. *Bureaucracy and Race: Native Administration in South Africa*. Berkeley: University of California Press.

Evans, Mei Mei. 2002. " 'Nature' and environmental justice." In Joni Adamson, Mei Mei Evans, and Rachel Stein, eds., *The Environmental Justice Reader: Politics, Poetics, and Pedagogy*. Tucson: University of Arizona Press, pp. 181–193.

Fanon, Franz. 1963. *The Wretched of the Earth*. New York: Grove Press.

Ferguson, James. 1990. *The Anti-Politics Machine: "Development," Depoliticization, and Bureaucratic Power in Lesotho*. Cambridge, UK: Cambridge University Press.

Ferguson, James. 1999. *Expectations of Modernity: Myths and Meanings of Urban Life on the Zambian Copperbelt*. Berkeley: University of California Press.

Ferguson, James. 2006. *Global Shadows: Africa and the Neo-Liberal World Ord*er. Durham, NC: Duke University Press.

Fiege, Mark. 1999. *Irrigated Eden: The Making of an Agricultural Landscape in the American West*. Seattle: University of Washington Press.

Fisher, Michael. 1973. *The Dam*. London: Constable.

Fontein, Joost. 2006. *The Silence of Great Zimbabwe: Contested Landscapes and the Power of Heritage*. London: University College London Press and Harare: Weaver.

Fortmann, Louise. 1995. "Talking claims: discursive strategies in contesting property." *World Development* 23(6):1053–1063.

Foster, Jeremy. 2008. *Washed with Sun: Landscape and the Making of White South Africa*. Pittsburgh: University of Pittsburgh Press.

Frederikse, Julie. 1982. *None but Ourselves: Masses vs. the Media in the Making of Zimbabwe*. Harare: Zimbabwe Publishing House.

Fredrickson, George M. 1981. *White Supremacy: A Comparative Study in American and South African History*. Oxford, UK: Oxford University Press.

Fuller, Alexandra. 2001. *Don't Let's Go to the Dogs Tonight*. New York: Random House.

Fuller, Alexandra. 2004. *Scribbling the Cat: Travels with an African Soldier*. New York: Penguin.

Funnekotter, A. n.d. "Facts about Kariba and Caribbea Bay." Memo for guests at the Caribbea Bay Hotel, Kariba, Zimbabwe.

Gande, Chris. 2005. *Section Eight*. Bloomington, IN: AuthorHouse.

Gaskill, Ivan. 1992. "History of images." In Peter Burke, ed., *New Perspectives in Historical Writing*. University Park: Pennsylvania State University Press, pp. 168–192.

Geertz, Clifford. 1963. *Agricultural Involution*. Berkeley: University of California Press.

Gerbi, Antonello. 1973. *The Dispute of the New World: The History of a Polemic, 1750–1900*. Translated by Jeremy Moyle. Pittsburgh: University of Pittsburgh Press.

Geschiere, Peter, and Francis Nyamnjoh. 2000. "Capitalism and autochthony: The seesaw of mobility and belonging." *Public Culture* 12(2):423–452.

Gifford, Paul. 1988. *The Religious Right in Southern Africa*. Harare: Baobab Books and University of Zimbabwe Publications.

Gillies, Colin. 1999. *Kariba in the Millenium, 1950–2000*. Bulawayo: Colin Gillies.

Gilroy, Paul. 2000. *Between Camps: Nations, Cultures, and the Allure of Race*. Cambridge, MA: Harvard University Press.

Glacken, Clarence J. 1967. *Traces on the Rhodian Shore: Nature and Culture in Western Thought from Ancient Times to the End of the Eighteenth Century*. Berkeley: University of California Press.

Godwin, Peter. 1996. *Mukiwa: A White Boy in Africa*. London: Picador.

Godwin, Peter. 2001. "Wildlife without borders: uniting Africa's wildlife reserves." *National Geographic* 200(3):2–29.

Godwin, Peter. 2006. *When a Crocodile Eats the Sun: A Memoir of Africa*. New York: Little, Brown.

Godwin, Peter, and Ian Hancock. 1993. *"Rhodesians never Die": the Impact of War and Political Change on White Rhodesia, c. 1970–1980*. Oxford, UK: Oxford University Press.

Gooder, Haydie, and Jane M. Jacobs. 2002. "Belonging and non-belonging: the apology in a reconciling nation." In Alison Blunt and Cheryl McEwan, eds., *Postcolonial Geographies*. London: Continuum, pp. 200–213.

Gordimer, Nadine. 1972. *The Conservationist*. London: Penguin.

Gordon, Robert. 1989. "Can Namibian San stop dispossession of their land." In Edwin N. Wilmsen, ed., *We Are Here: Politics of Aboriginal Land Tenure*. Berkeley: University of California Press, pp. 138–154.

Graf, William L. 2001. "*Dam*age control: restoring the physical integrity of America's rivers." *Annals of the Association of American Geographer* 91(1): 1–27.

Guha, Ramachandra. 1997. "The authoritarian biologist and the arrogance of anti-humanism: wildlife conservation in the Third World." *The Ecologist* 27(1):14–20.

Gumbo, Mafuranhunzi. 1995. *Guerrilla Snuff*. Harare: Baobab Books.

Hammar, Amanda, and Brian Raftopoulos. 2003. "Zimbabwe's unfinished business: rethinking land, state, and nation." In Hammar, Raftopoulos, and Jensen, eds., pp. 1–48.

Hammar, Amanda, Brian Raftopoulos, and Stig Jensen, eds. 2003. *Zimbabwe's Unfinished Business: Rethinking Land, State, and Nation in the Context of Crisis*. Harare: Weaver Press.

Hammond, Dorothy, and Alta Jablow. 1970. *The Africa That Never Was: Four Centuries of British Writing About Africa*. New York: Twayne Publishers.

Hanlon, Joseph. 1984. *Mozambique: The Revolution under Fire*. London: Zed.

Hansen, Karen Tranberg. 1989. *Distant Companions: Servants and their Employers in Zambia, 1900–1985*. Ithaca, NY: Cornell University Press.

Harris, Ashleigh. 2005. "Writing home: inscriptions of whiteness/descriptions of belonging in white Zimbabwean memoir-autobiography." In Muponde and Primorac, eds., pp. 103–117.

Harrison, Eric. 2006. *Jambanja*. Maioio Publishers.

Hart, David D., and Leroy Poff. 2002. "A special section of dam removal and river restoration." *BioScience* 52(8):653–655.

Hartz, Louis. 1964. *The Founding of New Societies*. New York: Harcourt, Brace and World.

Hellum, Anne, and Bill Derman. 2004. "Land reform and human rights in contemporary Zimbabwe: balancing individual and social justice through an integrated human rights framework." *World Development* 32(10):1785–1805.

Herbst, Jeffrey. 1988/89. "Racial reconciliation in southern Africa." *International Affairs* 65(1):43–54.

Herbst, Jeffrey. 2000. *States and Power in Africa: Comparative Lessons in Authority and Control.* Princeton, NJ: Princeton University Press.

Hill, Jane C. 1999. "Language, race, and white public space." *American Anthropologist* 100(3):680–689.

Hodder-Williams, Richard. 1983. *White Farmers in Rhodesia, 1890–1965: A History of the Marandellas District.* London: Macmillan.

Holding, Ian. 2005. *Unfeeling.* London: Scribner.

Holt, Thomas C. 2000. *The Problem of Race in the 21st Century.* Cambridge, MA: Harvard University Press.

Holtzclaw, Heather. 2004. "The third Chimurenga?: state terror and state organized violence in Zimbabwe's commercial farming communities." Doctoral dissertation, Michigan State University, East Lansing, Michigan.

hooks, bell. 2009. *Belonging: A Culture of Place.* New York: Routledge.

Horrell, Georgina. 2004. "A whiter shade of pale: white femininity as guilty masquerade in the 'new' (white) South African women's writing." *Journal of Southern African Studies* 30(4):765–776.

Hotchkiss, Jane. 1998. "Coming of age in Zambesia." In Elazar Barkan and Marie-Denis Shelton, eds., *Borders, Exiles, Diasporas.* Stanford, CA: Stanford University Press, pp. 81–91.

Houghton, John. n.d. "The planning of Kariba and the Zambezi Valley." Memo from the Mayor's Parlour, Municipality of Kariba, Zimbabwe.

Howarth, David. 1961. *The Shadow of the Dam.* New York: Macmillan.

Hughes, David McDermott. 2001. "Rezoned for business: how eco-tourism unlocked black farmland in eastern Zimbabwe." *Journal of Agrarian Change* 1(4):575–599.

Hughes, David McDermott. 2005. "Third nature: making space and time in the great Limpopo conservation area." *Cultural Anthropology* 20(2):157–184.

Hughes, David McDermott. 2006. *From Enslavement to Environmentalism: Politics on a Southern African Frontier.* Seattle: University of Washington Press and Harare: Weaver.

Hulme, David, and Marshall Murphree, eds. 2001. *African Wildlife and Livelihoods: The Promise and Performance of Community Conservation*. Oxford, UK: James Currey.

Huxley, Elspeth. 1959. *The Flame Trees of Thika: Memories of an African Childhood.* Harmondsworth, UK: Penguin.

Huxley, Elspeth. 1985. *Out in the Midday Sun*. Harmondsworth, UK: Penguin.

Isaacman, Allen, and Barbara Isaacman. 1975. "The prazeiros as transfrontiersmen: a study in social and cultural change." *International Journal of African Historical Studies* 8(1):1–39.

Ivy, P., and K. Bromley. 1972. *Marandellas Agricultural Survey*. Salisbury: Department of Conservation and Extension. Revised edition.

Jackson, Wes. 1994. *Becoming Native to This Place*. New York: Counterpoint.

Jacobs, Nancy. 2006. "The intimate politics of ornithology in colonial Africa." *Comparative Studies in Society and History* 48(3):564–603.

Jarosz, Lucy. 1992. "Constructing the dark continent: metaphor as geographic representation of Africa." *Geografiska Annaler* 74B(2): 105–115.

Jeater, Diana. 2001. "Speaking like a native: vernacular languages and the state in Southern Rhodesia, 1890–1935." *Journal of African History* 42:449–468.

Jeater, Diana. 2007. *Law, Language, and Science: The Invention of the "Native Mind" in Southern Rhodesia, 1890–1930*. Portsmouth, NH: Heinemann.

Johnson, R. W. 2000. "After the election." *London Review of Books*. July 20: 23–24.

Joubert, Leonie S. 2006. *Scorched: South Africa's Changing Climate*. Johannesburg: Wits University Press.

Jubb, R. A. 1961. *An Illustrated Guide to the Freshwater Fishes of the Zambezi River, Lake Kariba, Pungwe, Sabi, Lundi, and Limpopo Rivers*. Bulawayo, Zimbabwe: Stuart Manning.

Kann, Wendy. 2006. *Casting with a Fragile Thread: A Story of Sisters and Africa*. New York: Henry Holt.

Kenmuir, Dale. 1978. *A Wilderness Called Kariba: The Wildlife and Natural History of Lake Kariba*. Salisbury: Wilderness Publications.

Kenmuir, Dale. 1983. *Fishes of Kariba*. Harare: Wilderness Publications.

Kenmuir, Dale. 1987. *The Tusks and the Talisman*. Pretoria: De Jager-HAUM.

Kenmuir, Dale . 1990. *Dry Bones Rattling*. Pretoria: De Jager-HAUM.

Kenmuir, Dale. 1991. *Sing of Black Gold*. Pretoria: De Jager-HAUM.

Kenmuir, Dale. 1993. *The Catch*. Cape Town: Maskew Miller Longman.

Kennedy, Dane. 1987. *Islands of White: Settler Society and Culture in Kenya and Southern Rhodesia, 1890–1939*. Durham, NC: Duke University Press.

Kinsey, Bill H. 1999. "Land reform, growth, and equity: emerging evidence from Zimbabwe's resettlement program." *Journal of Southern African Studies* 25(2):173–196.

Kipling, Rudyard. 1928. *Rudyard Kipling's Verse, Inclusive Edition, 1885–1926*. Garden City, NY: Doubleday.

Kirkman, Penelope, ed. 2007. *A Gathered Radiance, Drawn from the Life of Martin van der Spuy: Letters, Paintings, and Drawings*. Harare: Penelope Kirkman.

Kopytoff, Igor. 1987. "The internal African frontier: the making of an African political culture." In Igor Kopytoff, ed., *The African Frontier: The Reproduction of Traditional African Societies*. Bloomington: Indiana University Press, pp. 3–84.

Kosek, Jake. 2006. *Understories: The Political Life of Forests in Northern New Mexico*. Durham, NC: Duke University Press.

Kramer, Eira. 2003. "Small-scale game conservancies." In Kramer, ed., pp. 11–17.

Kramer, Eira, ed. 2003. *Options for Wildlife on Zimbabwe's Highveld*. Harare: Department of Economic History, University of Zimbabwe.

Krog, Antjie. 1998. *Country of My Skull: Guilt, Sorrow, and the Limits of Forgiveness in the New South Africa*. Johannesburg: Random House South Africa.

Krog, Antjie. 2003. *A Change of Tongue*. Johannesburg: Random House South Africa.

Lagus, Charles. 1960. *Operation Noah*. New York: Coward-McCann.

Lamb, Christina. 2006. *House of Stone: The True Story of a Family Divided in War-Torn Zimbabwe*. Chicago: Lawrence Hill Books.

Langston, Nancy. 2003. *Where Land and Water Meet: A Western Landscape Transformed*. Seattle: University of Washington Press.

Lansing, J. Stephen. 1991. *Priests and Programmers: Technologies of Power in the Engineered Landscape of Bali*. Princeton, NJ: Princeton University Press.

Lefebvre, Henri. 1991. *The Production of Space*. (Donald Nicholson Smith, trans.) Oxford, UK: Blackwell.

Leicester Review of Books. 2007. "Interview: John Eppel, author of The Holy Innocents." February 4, 2007. Downloaded from http://leicesterreviewofbooks.wordpress.com on February 20, 2008.

Lemon, David. 1987. *Lake Kariba: Africa's Inland Sea*. Harare: Modus Publications.

Lemon, David. 1988. *Kariba Adventure*. Harare: College Press.

Lemon, David. 1997. *Hobo Rows Kariba*. Harare: African Publishing Group.

Lemon, David. 2000. *Never Quite a Soldier: A Policeman's War, 1971–1983*. Stroud, UK: Albida Press.

Lepore, Jill. 1998. *The Name of War: King Philip's War and the Origins of American Identity*. New York: Vintage.

Lessing, Doris. 1950. *The Grass Is Singing*. New York: Crowell.

Lessing, Doris. 1951. *African Stories*. New York: Ballantine Books.

Lessing, Doris. 1956. "The Kariba project." *New Statesman* 51 (June 9):647.

Lessing, Doris. 1958a. "Desert child." *New Statesman* 15 (November 15):700.

Lessing, Doris. 1958b. *Landlocked*. New York: Simon and Schuster.

Lessing, Doris. 1992. *African Laughter: Four Visits to Zimbabwe*. New York: HarperCollins.

Lessing, Doris. 1994. *Under My Skin: Volume One of my Autobiography, to 1949*. London: HarperCollins.

Lessing, Doris. 2003. "The Jewel of Africa." *The New York Review of Books* 50(6), April 10:6–10.

Li, Tania Murray. 2000. "Articulating indigenous identity in Indonesia: resource politics and the tribal slot." *Comparative Studies in Society and History* 42(1):149–179.

Li, Tania Murray. 2007. *The Will to Improve: Governmentality, Development, and Practice of Politics*. Durham, NC: Duke University Press.

Lim, Walter S. H. 1998. *The Arts of Empire: The Poetics of Colonialism: from Ralegh to Milton*. Newark, DE: University of Delaware Press.

Livingstone, David and Charles. 1865. *Narrative of an Expedition to the Zambesi and Its Tributaries*. London: John Murray.

Locke, John. 1980 [1690]. *Second Treatise of Government*. Edited by C. B. Macpherson. Indianapolis, IN: Hackett.

López, Alfred J. 2005. "Introduction: whiteness after empire." In Alfred J. López, ed. *Postcolonial Whiteness: A Critical Reader in Race and Empire*. Albany: State University of New York Press, pp. 1–30.

Louter, David. 2006. *Windshield Wilderness: Cars, Roads, and Nature in Washington's National Parks*. Seattle: University of Washington Press.

MacArthur, Robert, and Edward Wilson. 1967. *The Theory of Island Biogeography*. Princeton, NJ: Princeton University Press.

Macdonald, Sheila, ed. 2003. *Winter Cricket: The Spirit of Wedza*. Harare: Sheila Macdonald.

MacKenzie, John M. 1988. *The Empire of Nature: Hunting, Conservation and British Imperialism*. Manchester: Manchester University Press.

Magadza, C. H. D. 2000. "The distribution, ecology, and economic importance of lakes in Southern Africa." In M. J. Tumbare, ed. *Management of River Basins and Dams: The Zambezi River Basin.* Rotterdam: Balkema, pp. 283–295.

Magubane, Zine. 2004. *Bringing the Empire Home: Race and Class in Britain and Colonial South Africa.* Chicago: University of Chicago Press.

Main, Michael. 1987. *Kalahari: Life's Variety in Dune and Delta.* Johannesburg: Southern Book Publishers.

Main, Michael. 1990. *Zambezi: Journey of a River.* Halfway House, South Africa: Southern Book Publishers.

Malasha, Isaac. 2002. "Fisheries co-management: comparative analysis of the Zambian and Zimbabwean inshore fisheries of Lake Kariba." Ph.D. dissertation, University of Zimbabwe, Harare.

Malan, Rian. 1991. *My Traitor's Heart.* New York: Vintage.

Mamdani, Mahmood. 1996. *Citizen and Subject: Contemporary Africa and the Legacy of Late Colonialism.* Princeton, NJ: Princeton University Press.

Marciano, Francesca. 1999. *Rules of the Wild.* London: Vintage.

Marongwe, Nelson. 2003a. "Farm occupations and occupiers in the new politics of land in Zimbabwe." In Hammar, Raftopoulos, and Jensen, eds., pp. 155–190.

Marongwe, Nelson. 2003b. "Fast track resettlement and its implications for the wildlife land-use option: the case for Dahwye Resettlement Scheme." In Kramer, ed., pp. 18–26.

Martin, David. 1995. *Kariba: Nyaminyami's Kingdom.* Harare: African Publishing Group.

Marx, Leo. 1964. *The Machine in the Garden: Technology and the Pastoral Ideal in America.* Oxford, UK: Oxford University Press.

Mavedzenge, B. Z., J. Mahenehene, F. Murimbarimba, I. Scoones, and W. Wolmer. 2008. "The dynamics of real markets: cattle in southern Zimbabwe following land reform." *Development and Change* 39(4): 613–639.

Maxwell, David. 1999. *Christians and Chiefs in Zimbabwe: A Social History of the Hwesa People c. 1870s–1990s.* Edinburgh: Edinburgh University Press.

Matowanyika, Joseph Zano Zvapera. 1989. "Cast out of Eden: peasants versus wildlife policy in savanna Africa." *Alternatives* 16(1):30–39.

McClintock, Anne. 1995. *Imperial Leather: Race, Gender, and Sexuality in the Colonial Contest.* London: Routledge.

McClymont, D. S. 1981. *Bonanza: 75 Years of Flue-Cured Tobacco Advice.* Bulawayo, Zimbabwe: Books of Zimbabwe.

McCulloch, Jock. 2000. *Black Peril, White Virtue: Sexual Crime in Southern Rhodesia, 1902–1935*. Bloomington: Indiana University Press.

McCully, Patrick. 1996. *Silenced Rivers: The Ecology and Politics of Large Dams*. London: Zed.

McGregor, JoAnn. 1991. "Woodland resources: ecology, policy, and ideology—a historical case study of woodland use in Shurugwi Communal Area, Zimbabwe." D.Phil. dissertation, Loughborough University of Technology, UK.

McGregor, JoAnn. 2000. " 'The Great River': European and African images of the Zambezi." Paper presented to the "A View of the Land" conference, Bulawayo, July 3–7.

McGregor, JoAnn. 2003. "Living with the river: landscape and memory in the Zambezi Valley, northwest Zimbabwe." In William Beinart and JoAnn McGregor, eds., *Social History and African Environments*. Oxford, UK: James Currey, pp. 87–105.

McGregor, JoAnn. 2009. *Crossing the Zambezi: The Politics of Landscape on a Central African Frontier*. Woodbridge, UK: James Currey.

McKibben, Bill. 1989. *The End of Nature*. New York: Random House.

McPhee, John. 1971. *Encounters with the Archdruid*. New York: Farrar, Straus and Giroux.

McPhee, John. 1980. *Basin and Range*. New York: Farrar, Straus and Giroux.

Mda, Zakes. 2000. *The Heart of Redness*. Oxford, UK: Oxford University Press.

Meadows, Keith. 1981. *Rupert Fothergill: Bridging a Conservation Era*. Bulawayo: Thorntree Press.

Meadows, Keith. 1996. *Sand in the Wind*. Bulawayo: Thorntree Press.

Meadows, Keith. 2000. *Sometimes When It Rains: White Africans in Black Africa*. Bulawayo, Zimbabwe: Thorntree Press.

Memmi, Albert. 2000. *Racism*. Translated by Steve Martino. Minneapolis: University of Minnesota Press.

Mhanga, Alice T., R. D. Taylor, and R. J. Phelps. 1986. "HCH and DDT residues in the freshwater sardine (kapenta) at the Ume River mouth, Kariba." *Zimbabwe Science News* 20(3/4):46–49.

Mitchell, Don. 1996. *The Lie of the Land: Migrant Workers and the California Landscape*. Minneapolis: University of Minnesota Press.

Mitchell, Timothy. 1991. *Colonising Egypt*. Berkeley: University of California Press.

Moore, Donald S., Anand Pandian, and Jake Kosek. 2003. "The cultural politics of race and nature: terrains of power and practice," in Donald

S. Moore, Jake Kosek, and Anand Pandian, eds., 2003. *Race, Nature, and the Politics of Difference*. Durham, NC: Duke University Press, pp. 1–70.

Moore, Donald S., Jake Kosek, and Anand Pandian, eds. 2003. *Race, Nature, and the Politics of Difference*. Durham, NC: Duke University Press.

Moore, J. W. H. 1965. "A dam on the Zambezi." *National Archives of Rhodesia Occasional Papers* 1:41–59.

Moore-King, Bruce. 1988. *White Man, Black War*. Harare: Baobab.

Moreau, Jacques, ed. 1997. *Advances in the Ecology of Lake Kariba*. Harare: University of Zimbabwe Press.

Moyana, Henry V. 1984. *The Political Economy of Land in Zimbabwe*. Gweru, Zimbabwe: Mambo Press.

Moyo, Sam. 1995. *The Land Question in Zimbabwe*. Harare: Southern African Regional Institute for Policy Studies.

Moyo, Sam. 2000. *Land Reform under Structural Adjustment in Zimbabwe: Land Use Change in the Mashonaland Provinces*. Uppsala, Sweden: Nordiska Afrikainstitutet.

Moyo, Sam. 2005. "Land occupations and land reform in Zimbabwe: towards the national democratic revolution." In Sam Moyo and Paris Yeros, eds., *Reclaiming the Land: The Resurgence of Rural Movements in Africa, Asia, and Latin America*. London: Zed, pp. 165–205.

Muponde, Robert, and Ranka Primorac, eds. 2005. *Versions of Zimbabwe: New Approaches to Literature and Culture*. Harare: Weaver Press.

Murphree, Marshall W. 1978. "Whites in black Africa: their status and role." *Ethnic and Racial Studies* 1(2):154–174.

Murphy, Ian. n.d. "Kariba: Africa's best kept secret." Kariba: Kariba Publicity Association.

Murray, Barbara. 1995. "Jean Hahn: impressionist of Africa." *Gallery* (National Gallery of Zimbabwe) 5:9–11.

Muzondidya, James. 2007. "*Jambanja*: ideological ambiguities in the politics of land and resource ownership in Zimbabwe." *Journal of Southern African Studies* 33(2):325–341.

Myers, Garth Andrew. 2003. *Verandahs of Power: Colonialism and Space in Urban Africa*. Syracuse, NY: Syracuse University Press.

Naipaul, Shiva. 1980. *North of South: An African Journey*. New York: Penguin.

Nash, Roderick. 1967. *Wilderness and the American Mind*. New Haven, CT: Yale University Press.

National Federation of Women's Institutes. 1967. *Great Spaces Washed with Sun*. Salisbury: M. O. Collins.

Neumann, Roderick. 1998. *Imposing Wilderness: Struggles over Nature and Livelihood in Africa*. Berkeley: University of California Press.

Nicolson, Marjorie Hope. 1959. *Mountain Gloom and Mountain Glory: The Development of the Aesthetics of the Infinite*. Ithaca, NY: Cornell University Press.

Nixon, Rob. 1999. *Dreambirds: The Strange History of the Ostrich in Fashion, Food, and Fortune*. London: Doubleday.

Nuttall, Sarah. 1996. "Flatness and fantasy: representations of the land in two recent South African novels." In Kate Darian-Smith, Liz Gunner, and Sarah Nuttall, eds., *Text, Theory, Space: Land, Literature, and History in South Africa and Australia*. London: Routledge, pp. 219–230.

Nye, David E. 1994. *American Technological Sublime*. Cambridge, MA: MIT Press.

Nyschens, Ian. 1997. *Months of the Sun*. Long Beach, CA: Safari Press.

Ong, Aihwa. 1999. *Flexible Citizenship: The Cultural Logics of Transnationality*. Durham, NC: Duke University Press.

Orlove, Benjamin S. 1996. "Struggles to control the commons: social movement or cultural emplacement?" Paper presented to the International Association for the Study of Common Property, Berkeley, CA, June 5–9.

Orlove, Benjamin S. 2002. *Lines in the Water: Nature and Culture at Lake Titicaca*. Berkeley: University of California Press.

Ostrom, Elinor. 1990. *Governing the Commons: The Evolution of Institutions for Collective Management*. Cambridge, UK: Cambridge University Press.

Palmer, Robin. 1977. *Land and Racial Domination in Rhodesia*. Berkeley: University of California Press.

Palmer, Robin. 1990. "Land reform in Zimbabwe, 1980–1990." *African Affairs* 89(355):163–181.

Pape, John. 1990. "Black and white: the 'perils of sex' in colonial Zimbabwe." *Journal of Southern African Studies* 16(4):699–720.

Passaportis, Michael Steven. 2000. "Resistance and the post-colonial state: the case of white Zimbabwean farmers." Bachelor's thesis, Harvard University, Cambridge, MA.

Peach, Margaret. 2003. *My Place in the Sun: An Insight into the Life of a Game Warden's Wife in Rhodesia during the Period 1960–1980*. London: Polyphrenos Limited.

Peek, Bookey. 2007. *All the Way Home: Stories from an African Wildlife Sanctuary*. London: Little Books.

Phimister, Ian. 1989. "Discourse and the discipline of historical context: conservationism and ideas about development in southern

Rhodesia, 1930–1950." *Journal of Southern African Studies* 12(2): 263–275.

Pitman, Dick. 1979. *You Must Be New Around Here*. Bulawayo: Books of Rhodesia.

Pitman, Dick. 1980. *Wild Places of Zimbabwe*. Bulawayo: Books of Zimbabwe.

Pitman, Dick. 1983a. "The mighty Zambezi: part two, the inland sea." *Africa Calls from Zimbabwe* 140:10–18.

Pitman, Dick. 1983b. "The Zambezi group." *Zimbabwe Wildlife* 32:10.

Pitman, Dick. 2008. *A Wild Life: Adventures of an Accidental Conservationist in Africa*. Guilford, CT: Lyons Press.

Pitman, Dick. n.d. *Zimbabwe Portrait*. Harare: Modus Publications.

Pollan, Michael. 1991. *Second Nature: A Gardener's Education*. New York: Dell.

Potgieter, Herman. 1990. *South Africa: Landshapes, Landscapes, Manshapes*. (Text by Guy Butler.) Cape Town: Struik.

Pratt, Mary Louise. 1992. *Imperial Eyes: Travel Writing and Transculturation*. London: Routledge.

Pred, Allan. 1986. *Place, Practice, and Structure: Social and Spatial Transformation in Southern Sweden, 1750–1850*. Cambridge, UK: Polity Press.

Price, A. Grenfell. 1939. *White Settlers in the Tropics*. New York: American Geographical Society.

Primorac, Ranka. 2006. *The Place of Tears: The Novel and Politics in Modern Zimbabwe*. London: I. B. Tauris.

Rabinow, Paul. 1989. *French Modern: Norms and Forms of the Social Environment*. Cambridge, MA: MIT Press.

Raffles, Hugh. 2002. *In Amazonia: A Natural History*. Princeton, NJ: Princeton University Press.

Raftopoulos, Brian. 2005. "Unreconciled difference: the limits of reconciliation politics in Zimbabwe." In Raftopoulos and Savage, eds., pp. viii–xxii.

Raftopoulos, Brian, and Tyrone Savage, eds. 2005. *Zimbabwe: Injustice and Political Reconciliation*. Cape Town: Institute for Justice and Reconciliation.

Ranger, Terence. 1996. "'Great spaces washed with sun': the Matopos and Uluru compared." In Kate Darian-Smith, Liz Gunner, and Sarah Nuttall, eds., *Text, Theory, Space: Land, Literature, and History in South Africa and Australia*. London: Routledge, p. 159.

Ranger, Terence. 1999. *Voices from the Rocks: Nature, Culture and History in the Matopos Hills of Zimbabwe*. Oxford, UK: James Currey.

Ranger, Terence. 2004. "Nationalist historiography, patriotic history and the history of the nation: the struggle over the past in Zimbabwe." *Journal of Southern African Studies* 30(2):215–234.

Ravenstein, E. G. 1891. "Lands of the globe still available for European settlement." *Proceedings of the Royal Geographical Society* 13(1):27–35.

Rayner, Richard. 1980. *The Valley of Tantalika: An African Wild Life Story*. Bulawayo: Books of Zimbabwe.

Read, Peter. 2000. *Belonging: Australians, Place, and Aboriginal Ownership*. Cambridge, UK: Cambridge University Press.

Redfield, Peter. 2000. *Space in the Tropics: From Convicts to Rockets in French Guiana*. Berkeley: University of California Press.

Reeve, W. H. 1960. "Progess and geographical significance of the Kariba dam." *Geographical Journal* 126(2):140–146.

Reynolds, Henry. 1998. *This Whispering in our Hearts*. St. Leonards, Australia: Allen and Unwin.

Reynolds, Pamela. 1991. *Dance Civet Cat: Child Labour in the Zambezi Valley*. Harare: Baobab and Athens, OH: Ohio University Press.

Reynolds, R. 1964. "The British South Africa Company's central settlement farm, Marandellas, 1907–1910 from the papers of H. K. Scorror." *Rhodesiana* 10:1–16.

Reynolds, R. 1978. "New island resort at Kariba." *Rhodesia Calls* 108:9–12.

Reynolds, R. 1969. "Rhodesia's lakes." *Rhodesia Calls* 56:8–11.

Robins, Eric, and Ronald Legge. 1959. *Animal Dunkirk: The Story of Kariba Dam*. New York: Taplinger.

Roediger, David R. 1991. *Wages of Whiteness: Race and the Making of the American Working Class*. New York: Verso.

Roediger, David R. 2002. *Colored White: Transcending the Racial Past.* Berkeley: University of California Press.

Rogers, David. 2005. *Zambia Safari in Style*. Cape Town: Africa Geographic.

Rothchild, Donald S. 1973. *Racial Bargaining in Independent Kenya: A Study of Minorities and Decolonization*. London: Oxford University Press.

Roux, François. 1996. "Clarity and discipline: an appreciation of Robert Paul." In Barbara Murray, ed. *Robert Paul.* Harare: Colette Wiles, pp. 58–63.

Ruark, Robert. 1962. *Uhuru*. London: Corgi.

Rubert, Steven C. 1998. *A Most Promising Weed: A History of Tobacco Farming and Labor in Colonial Zimbabwe, 1890–1945.* Athens, OH: University of Ohio Press.

Rugube, Lovemore, Sam Zhou, Michael Roth, and Walter Chambati. 2003. "Government-assisted and market-driven land reform: evaluating public and private land markets in redistributing land in Zimbabwe." In Michael Roth and Francis Gonese, eds., *Delivering Land and Securing Rural Livelihoods: Post-Independence Land Reform and Resettlement in Zimbabwe.* Harare: Centre for Applied Social Sciences, University of Zimbabwe, pp. 115–139.

Rukuni, Mandivamba. 1994. "The evolution of agricultural policy, 1890–1990." In Mandivamba Rukuni and Carl K. Eicher, eds., *Zimbabwe's Agricultural Revolution*. Harare: University of Zimbabwe Press, pp. 15–39.

Rukuni, Mandivamba, ed. 1994. *Report of the Commission of Enquiry into Appropriate Agricultural Land Tenure Systems*. 3 vols. Harare: Government Printer.

Rutherford, Blair. 2001. *Working on the Margins: Black Workers, White Farmers in Postcolonial Zimbabwe*. London: Zed and Harare: Weaver.

Rutherford, Blair. 2003. "Belonging to the farm(er): farm workers, farmers, and the shifting politics of citizenship." In Hammar, Raftopoulos, and Jensen, eds., pp. 191–216.

Sachikonye, Lloyd M. 2005. "The promised land: from expropriation to reconciliation and *jambanja*." In Raftopoulos and Savage, eds., pp. 1–18.

Said, Edward W. 1993. *Culture and Imperialism*. New York: Knopf.

Schama, Simon. 1995. *Landscape and Memory*. New York: Knopf.

Schroeder, Richard. 1999. "Geographies of environmental intervention in Africa." *Progress in Human Geography* 23(3):359–378.

Schroeder, Richard. 2008. "South African capital in the land of Ujamaa: contested terrain in Tanzania." *African Sociological Review* 12(1):20–34.

Schutz, Barry. 1972. *The Theory of Fragment and the Political Development of White Settler Society in Rhodesia*. Berkeley: University of California Press.

Schwab, Glenn O., Delmar O. Fangmeier, William J. Elliot, and Richard K. Prevert. 1993. *Soil and Water Conservation Engineering*. New York: Wiley.

Scudder, Thayer. 1962. *The Ecology of the Gwembe Tonga*. Manchester: Manchester University Press.

Seddon, George. 1997. *Landprints: Reflections on Place and Landscape*. Cambridge University Press, Cambridge.

Selby, Angus. 2006. "Commercial farmers and the state: interest group politics and land reform in Zimbabwe." Ph.D. dissertation, Oxford University, Oxford, UK.

Shutt, Allison K. 2002. "The settlers' cattle complex: the etiquette of culling cattle in colonial Zimbabwe, 1938." *The Journal of African History* 43(2):263–286.

Skarpe, Christina. 1997. "Ecology of the vegetation in the draw-down zone of Lake Kariba." In Moreau, ed., pp. 120–138.

Smith, Ian. 2001. *Bitter Harvest: The Great Betrayal and the Dreadful Aftermath*. Johannesburg: Jonathan Ball.

Smithers, Reay. 1959. "The Kariba lake." *Oryx* 5(1):21–24.

Soils Incorporated. 2000. *Kariba Dam: Zambia and Zimbabwe*. Cape Town: World Commission on Dams.

Solnit, Rebecca. 1994. *Savage Dreams: A Journey into the Hidden Wars of the American West*. San Francisco: Sierra Club Books.

St. John, Lauren. 2007. *Rainbow's End: A Memoir, of Childhood, War and an African Farm*. New York: Scribner.

St. Leger, Gavin. 2004. *River Road*. (Photographic essay on compact disc).

Staunton, Irene, ed. 2005. "Damage: the personal costs of political change in Zimbabwe." Unpublished ms.

Stegner, Wallace. 1992. *Where the Bluebird Sings to the Lemonade Springs: Living and Writing in the West*. New York: Random House.

Steinberg, Johnny. 2002. *Midlands*. Johannesburg: Jonathan Ball.

Steyn, Melissa. 2001. *"Whiteness just isn't what it used to be: White Identity in a Changing South Africa*. Albany: State University of New York Press.

Stockley, Cynthia. 1911. *The Claw: A Story of South Africa*. New York: Grosset and Dunlap.

Stoler, Ann L. 1989. "Making empire respectable: the politics of race and sexual morality in 20th-century colonial cultures." *American Ethnologist* 16(4):634–660.

Stoler, Ann L. 2002. *Carnal Knowledge and Imperial Power: Race and the Intimate in Colonial Rule*. Berkeley: University of California Press.

Stoler, Ann L. 2004. "Affective states." In David Nugent and Joan Vincent, eds., *A Companion to the Anthropology of Politics*. Oxford, UK: Blackwell, pp. 4–20.

Struthers, John. 1993. Life and Death of a Pool. Shreusbury, UK: Swan Hill Press.

Stutchbury, Jeff, and Veronica. 1992. *Spirit of the Zambezi*. London: CBC Publishing.

Suzuki, Yuka. 2007. "Putting the lion out at night: domestication and the taming of the wild." In Rebecca Cassidy and Molly Mullin, eds., *Where the Wild Things Now Are: Domestication Reconsidered*. Oxford, UK: Berg, pp. 229–247.

Swyngedouw, Eric. 1999. "Modernity and hybridity: nature, *regeneracionismo*, and the production of the Spanish waterscape, 1890–1930." *Annals of the Association of American Geographers* 89(3): 443–465.
Taylor, Bill. 2002. *Wet Breams*. Bulawayo: Sheltom.
Taylor, Julie. 2002. "The politics of uncertainty in a white Zimbabwean farming community." Bachelor's thesis, Cambridge University, Cambridge, UK.
Taylor, Rex. 1993. "Kariba." *Zimbabwe Fisherman* 5(5):15–16.
Taylor, Rex. 1995. "Kariba." *Zimbabwe Fisherman* 7(4):33.
Taylor, Rex. 1997. "A sailing vessel for Lake Kariba: the Zeemin." *Zimbabwe Fisherman* 9(2):26.
Taylor, Rex. 1999. "Kariba." *Zimbabwe Fisherman* 11(1):40.
Teede, Jan and Fiona. 1990. *The Zambezi: River of the Gods*. London: Andre Deutsch, South Africa: Russel Friedman Books, and Harare: Ruscombe Books.
Theroux, Paul. 1983. *The Kingdom by the Sea: A Journey Around Great Britain*. Boston: Houghton Mifflin.
Thomas, Keith. 1983. *Man and the Natural World: A History of the Modern Sensibility*. New York: Pantheon.
Thompson, E.P. 1963. *The Making of the English Working Class*. London: V. Gollancz.
Timberlake, Jonathan. 1998. *Biodiversity of the Zambezi Basin Wetlands, Phase 1: Review and Preliminary Assessment of Available Information.* Vol. 1. Harare: Zambezi Society.
Todd, Judith. 2007. *Through the Darkness: A Life in Zimbabwe*. Cape Town: Zebra Press.
Urry, John. 1990. *The Tourist Gaze: Leisure and Travel in Contemporary Societies*. London: Sage.
Uusihakala, Katja. 1999. "From impulsive adventure to postcolonial commitment: making white identity in contemporary Kenya." *European Journal of Cultural Studies* 2(1):27–45.
Venables, Bernard. 1974. *Coming Down in the Zambezi*. London: Constable.
Viertel, Peter. 1960. *White Hunter, Black Heart*. London: Panther.
Wagner, Kathrin. 1994. *Rereading Nadine Gordimer*. Johannesburg: Witwatersrand University Press and Bloomington: Indiana University Press.
Wainaina, Binyavanga. 2006. "How to write about Africa." *Granta* issue 92.
Wallace, Anthony F. C. 1999. *Jefferson and the Indians: The Tragic Fate of the First Americans*. Cambridge, MA: Harvard University Press.
Walsh, Kevin, and Anthony Williams. 1993. *The Fisherman's Guide to Zimbabwe*. Harare: Mag-Set Publications.

Wannenburgh, Alf. 1978. *Rhodesian Legacy.* (Photos by Ian Murphy) Cape Town: Struik.

Ward, David. 1989. *Chronicles of Darkness.* London: Routledge.

Watt, Duncan. 1992. *Killers Against Kariba.* Lutterworth, UK: Tynron Press.

Waugh, Evelyn. 1960. *A Tourist in Africa.* Boston: Little, Brown.

Weinrich, A. K. H. 1975. *African Farmers in Rhodesia: Old and New Peasant Communities in Karangaland.* London: Oxford University Press.

White, Luise. 2004. "Precarious conditions: a note on counter-insurgency in Africa after 1945." *Gender and History* 16(3):603–625.

White, Richard. 1991. *The Middle Ground: Indians, Empires, and Republics in the Great Lakes Region, 1650–1815.* Cambridge, UK: Cambridge University Press.

White, Richard. 1995. *The Organic Machine: The Remaking of the Columbia River.* New York: Hill and Wang.

Whitlow, R. 1988. "Aerial photography in Zimbabwe, 1935–1986." *Zambezia* 15(2):137–165.

Wild, H., and A. Fernandes. 1968. *Flora Zambesiaca Supplement: Vegetation Map.* Salisbury: M. O. Collins.

Wilder, George A. 1933. *The White African.* Bloomfield, NJ: Morse Press.

Wildman, Stephanie M., and Adrienne D. Davis. 2000. "Language and silence: making systems of privilege visible." In Richard Delgado and Jean Stefancic, eds., *Critical Race Theory: The Cutting Edge.* 2nd ed. Philadelphia: Temple University Press, pp. 657–663.

Wiles, Richard F. 2005. *Foredoomed Is My Forest: The Diary of a Zimbabwe Farmer.* Victoria, Canada: Trafford Publishing.

Williams, Anthony. 1992. "Amanzi Africa." *The Zimbabwe Fisherman* 4(4):6, 35.

Williams, Anthony. 1997. "Chete Island, Westlake Safaris." *The Zimbabwe Fisherman* 9(6):10–11.

Williams, Anthony. 1998a. "Hungwe Lodge, Msuna." *The Zimbabwe Fisherman* 10(3):28–29.

Williams, Anthony. 1998b. "Editor's comment." *The Zimbabwe Fisherman* 10(5):3.

Williams, Anthony. 1998c. "Ode to trees." *The Zimbabwe Fisherman* 10(supplement):1.

Williams, Anthony. 1999a. "Editor's comment." *The Zimbabwe Fisherman* 11(5):3.

Williams, Anthony. 1999b. "Editor's comment." *The Zimbabwe Fisherman* 11(6):3.

Williams, Anthony. 1999c. "Environmental spotlight." *The Zimbabwe Fisherman* 11(4):22

Williams, Anthony. 2000. "Editor's Comment." *African Fisherman* 12(6):3.

Williams, Anthony. 2001a. "Report illegal fishing at Kariba." *African Fisherman* 13(5):20.

Williams, Anthony. 2001b "Anti-poaching task force launched." *African Fisherman* 13(2):42–43.

Williams, Anthony. 2002. "Fothergill Island—legendary hospitality." *African Fisherman* 14(6):6–8.

Williams, Charles. 1979. "New wilderness safari at Kariba." *Rhodesia Calls* 114:10–17.

Williams, Raymond. 1973. *The County and the City*. Oxford, UK: Oxford University Press.

Williams, Raymond. 1977. *Marxism and Literature*. Oxford, UK: Oxford University Press.

Wilson, Alexander. 1991. *The Culture of Nature: North American Landscape from Disney to the Exxon Valdez*. Toronto: Between the Lines.

Wilson, K. B. 1989. "Trees in fields in southern Zimbabwe." *Journal of Southern African Studies* 15(2):369–383.

Wolmer, W. L. 2001. "Lowveld Landscapes: Conservation, Development, and the Wilderness Vision in South-eastern Zimbabwe." Ph.D. dissertation, University of Sussex.

Wolmer, W. L. 2007. *From Wilderness Vision to Farm Invasions: Conservation and Development in Zimbabwe's South-East Lowveld*. Oxford, UK: James Currey.

Wood, L. 2003. "Report on irrigation dams in the Macheke/Virginia ICA as at December 1999." Unpublished document.

Worby, Eric. 2003. " 'The end of modernity in Zimbabwe: passages from development to soveignty." In Hammar, Raftopoulos, and Jensen, eds., pp. 49–82.

Wordsworth, William. 1991[1822]. *A Description of the Lakes*. Oxford, UK: Woodstock Books.

World Commission on Dams. 2000. "Kariba Dam, Zambia and Zimbabwe." Cape Town: World Commission on Dams.

Wynn, Sally. 1998. "Perceptions of wilderness: some preliminary observations from the Zambezi Society 1998 wilderness survey." Proceedings of the Fifth Seminar Held at Mavuradona Wilderness Camp, August 9–12, 1998. Harare: Zambezi Society.

Wynn, Sally. 2000. "The Zambezi River: wilderness and tourism." Harare: Zambezi Society.

Zimbabwe Fisherman. 1999. "Kariba International Tiger Fishing Tournament" (video). Harare: Zimbabwe Fisherman.

Zimmerer, Karl. 2000. "The reworking of conservation geographies: non-equilibrium landscapes and nature-society hybrids." *Annals of the Association of American Geographers* 99(2):356–369.

Index

Note: page numbers in *italics* refer to illustrations and the letter 'n' followed by locators denotes note numbers.